ALMATY–ISSYK-KUL ALTERNATIVE ROAD
ECONOMIC IMPACT ASSESSMENT

DECEMBER 2020

Economic Corridor Almaty–Bishkek

CAREC
Central Asia Regional Economic Cooperation Program

ADB

ISBN 978-92-9262-642-6 (print), 978-92-9262-643-3 (electronic), 978-92-9262-644-0 (ebook)
Publication Stock No. TCS200423-2
DOI: http://dx.doi.org/10.22617/TCS200423-2

Notes:
In this publication, "$" refers to United States dollars.
ADB recognizes "Kyrgyzstan" as the Kyrgyz Republic.
All photos are from the consultant team unless mentioned otherwise.

On the cover: **Late season at Lake Issyk-Kul.** A beach view from Cholpon-Ata in the Kyrgyz Republic (photo by Mark Sieber).

Contents

Tables, Figures, Box, and Maps

Box

Maps

Acknowledgments

This report was prepared by a team of specialists headed by Mark Sieber from the Boston and Zurich offices of EBP (formerly Economic Development Research Group), with local support from ILF Kazakhstan LLC as well as Elvira Ennazarova (Kyrgyz Republic) and Akhmadjon Niyozov (Kazakhstan).

The Almaty–Bishkek Economic Corridor team of Kristian Rosbach, Carmela Espina, Kylychbek Djakypov, Almas Baitenov, and Aidana Berdybekova provided guidance and support during the preparation of this report.

All photos are from the consultant team, unless specified otherwise.

Abbreviations

ABEC — Almaty–Bishkek Economic Corridor
ADB — Asian Development Bank
BCP — border crossing point
CBA — cost–benefit analysis
EBRD — European Bank for Reconstruction and Development
EIA — economic impact assessment
EIRR — economic internal rate of return
GDP — gross domestic product
MRIO — multiregional input–output
NPV — net present value
PPP — public–private partnership
ROI — return on investment
UN — United Nations

Executive Summary

For decades, there have been plans for a more direct road connection between Almaty and Issyk-Kul. Almaty, a vibrant city of 1.9 million in Kazakhstan, is only 80 kilometers away from Cholpon-Ata, one of the centers of the Issyk-Kul tourist destination popular for its lake, mountains, and moderate summers. Issyk-Kul in the Kyrgyz Republic, separated from Almaty by two chains of mountains, is home to half a million residents. The trip from Almaty to Cholpon-Ata or vice versa bypasses the mountains, measures 460 kilometers, and takes approximately 6.5 hours by passenger car and considerably more for regular buses. The border crossing points add additional (inestimable) time to the journey.

The Asian Development Bank (ADB) is exploring the potential for economic impacts that an alternative road would have. Within the framework of the Almaty–Bishkek Economic Corridor (ABEC*), this economic impact assessment (EIA) identifies the likely economic impacts on both sides of the border. By improving travel and lowering barriers to travel for regional and international tourists, the project aims at transnational economic development by expanding tourism demand. Other travelers who are not tourists would also benefit from lower travel times.

Issyk-Kul, a major tourism destination with more than 3.5 million visitors per year, has the potential to become a regional tourism magnet, according to the ABEC Tourism Master Plan (2019). An alternative road between Almaty and Issyk-Kul would remove a major obstacle to economic development. An easier and more potent flow of tourists into Issyk-Kul would lead to investments in the tourism industry in terms of capacity and quality.

Almaty, as a vibrant economic business center, has the tourist infrastructure to welcome more travelers. Almaty and its surrounding tourist sites could play a more important role as a destination for international travelers and would also see more Kyrgyz visitors from Issyk-Kul oblast. Almaty International Airport could strengthen its role as an important air hub by bringing additional international travelers to the region.

Several possible alignments for the new road are considered in this study (Chapter 3), with a wide range of resulting construction costs and travel times. They represent the variety of potential options for a new road. Among the considered alignments, construction costs are estimated to range from $81 million to $587 million.

Travel time will range from 1 hour and 40 minutes to just over 5 hours with the new road. The shorter travel time will enable people to travel more frequently and it will attract new visitor market segments, like weekend trippers. Those who already are traveling in the base case will benefit from the more direct route by experiencing travel time and cost savings. Those who choose to travel as a result of the much shorter travel times will add to spending and economic activity at their destination on either side of the border.

---

* ABEC comprises the two cities of Almaty and Bishkek, Almaty and Zhambyl oblasts on the Kazakh side of the border, and Chuy and Issyk-Kul oblasts on the Kyrgyz side.

Crossing the high mountain ridges between Almaty and Issyk-Kul is not possible in winter. Only one of the chosen alignments—effectively bypassing the highest mountains, climbing to moderate altitudes of 2,000 meters, and capping the ascent and descent through a tunnel—could potentially be kept open year-round. For all other alignments, it would be extremely costly to construct a road protected from snow and avalanches.

Additional spending is expected to justify investments in tourism and other infrastructure, not only in Issyk-Kul but also in the Almaty oblast. To exploit the full potential of a new road, investments are necessary in quality of services and capacity to serve visitors. An enhanced tourism product on both sides of the border will contribute to a further increase in tourism demand.

While shorter travel times and lower travel costs are seen as a key precondition for increased tourism demand and economic development, the alternative road is not sufficient on its own. Two scenarios taken from the ABEC Tourism Master Plan (scenarios II and III) are used to describe the degree to which additional policies and investments in tourism infrastructure are needed to further support growth in tourism (Chapter 4). The travel demand estimates were developed based on proven methodologies and linked to the respective tourism forecasts from the Tourism Master Plan.

The travel demand estimates are qualitatively underpinned by interviews conducted with individuals and tourism stakeholders in Almaty city, Issyk-Kul oblast, and Bishkek. They show a clear desire of Almaty residents to travel more frequently to Issyk-Kul if travel times were shorter. Weekend trips from Almaty seem to be not well-known. While Almaty has popular destinations for day trips in its immediate surroundings, it does not support weekend trips to other tourism destinations farther away, because of long travel times and limited supporting infrastructure for overnight trips away from the city. An alternative road between Almaty and Issyk-Kul will, to different extents depending on the alignment, open up new travel markets.

At the same time, the region will become more attractive for international travelers, as it will be easier to get around and combine multiple destinations in and around Almaty and Issyk-Kul. From the perspective of international travelers, Almaty and Issyk-Kul and their surroundings may be seen as a single mountain and lake tourism cluster.

Average daily traffic is estimated to range from 1,000 to 2,750 vehicles per day from June to November under scenario II conditions, and from 1,800 to 6,800 vehicles per day under scenario III conditions.

Those traveling between Almaty and Issyk-Kul will experience benefits from shorter travel times and lower out-of-pocket costs. The yearly savings are expected to range from $4 million to $23 million, depending on the alignment and the policy scenario (Chapter 5).

A multiregional input–output (MRIO) economic model was built for the purpose of this study. The MRIO economic model was used to determine direct, indirect, and induced economic impacts of additional economic activity.[**] As the ABEC regional economies are intertwined already, actual data about trade between the two countries was included in the MRIO economic model. The data revealed that every additional dollar spent in Issyk-Kul has a positive spillover impact of a few cents on the economy of Kazakhstan, as businesses in Issyk-Kul seem to depend on purchases of additional goods and services not available in their own country. Examples are food and agricultural goods, as well as gasoline. This is less the case going the other way. As Kazakhstan is

---

[**] For example, while additional revenue for businesses in tourism represents direct impacts, additional purchases paid for by that additional revenue are indirect impacts (e.g., agricultural products purchased to produce meals for additional tourists). Businesses need a larger workforce to serve additional tourists, generating more income. That additional income spent on purchases is counted as induced impacts.

the larger and less dependent economy, a larger share of the additional money spent in Almaty city and Almaty oblast remains in Kazakhstan.

Economic impacts as a consequence of increased economic activity enabled by the alternative road are considerable for both countries. On average, $31 million–$165 million will be added each year to the national gross domestic product (GDP) of Kazakhstan, depending on the alignment and policy scenario. This will range from $53 million to $439 million per year in the Kyrgyz Republic, which corresponds to 0.8%–6.7% of the 2017 Kyrgyz national GDP.

The impacts on employment are a considerable aspect of economic development enabled by the alternative road. About half of the additional jobs in the Kyrgyz Republic and 25%–33% in Kazakhstan will be in the hospitality industry. It may be a development constraint to fill that many new jobs in tourism with trained workers, especially in the Kyrgyz Republic, unless efforts are made to further invest in workforce training and housing.

Additional revenue for businesses and new income for workers could potentially increase the tax base in both countries. The western alignment through New Kastek Pass under scenario III would increase the tax revenue in Kazakhstan by $9 million and in the Kyrgyz Republic by $18 million. Depending on increased economic output, tax revenues from other alignment alternatives would be higher or lower.

A comprehensive understanding of economic internal rate of return (EIRR) was used to include broader economic development in the region as the primary goal of the road, measured by GDP growth. This EIRR exceeds ADB's 9% threshold for economic viability for all alignments in both policy scenarios. However, all alignments require travel demand to increase considerably, which, in light of the alternative road's advantages and the existing latent demand suggested by interviewees, is very likely (Chapter 6).

An approximation of the financial internal rate of return was conducted too. Based on assumptions for a tolling scheme, not all alignments prove to be able to generate sufficient toll revenue to cover the cost for operation and maintenance and of the tolling itself. However, only a comprehensive financial analysis can compare various ways to generate revenue to recover this cost and determinate the optimal toll rate that covers most of the cost while not deterring too many travelers.

Various ways of splitting the cost between the two countries are laid out in this report. A simple split by country could lead to very different results depending on the alignment. It could also be seen as unfair, as benefits and economic impacts from the road are unevenly distributed between the two countries. Cost split considerations could include the respective benefits and economic impacts occurring on each side of the border. A potential binational public–private partnership (PPP) scheme could reflect such considerations.

While economic development and its impacts are at the center of this study, other effects of an alternative road between Almaty and Issyk-Kul are not ignored. A multi-criteria rating considers that the new road could negatively impact national parks and other natural resources, but could add value to the experiences of travelers crossing the mountain area at high altitudes with glaciers and impressive scenery. These effects will depend on specific road alignment details to be determined later.

Two different basic alignment choices seem possible:

(i)   An alternative road at moderate cost (represented by the western alignments without a tunnel) promises to be more independent from strong increases in travel demand. Even though these alignments offer limited travel time and cost savings, they show the strongest economic and financial viability.

(ii)   A more direct alternative road (especially a direct alignment between Almaty and Cholpon-Ata) offering significant travel time and cost savings requires considerably higher capital investments and is a riskier endeavor. If supported by effective policies and projects as supposed in scenario III of the ABEC Tourism Master Plan, this choice promises stronger economic development.

It is the purpose of this study to demonstrate the economic impacts an alternative road between Almaty and Issyk-Kul would have for Kazakhstan and the Kyrgyz Republic. Ignoring the differences between the alignments and the policy scenarios, a positive economic impact on the economy of the ABEC region and both national economies can be clearly stated. There are economically viable solutions that, in a supporting policy environment, enable potentially strong economic development in the region. Those solutions are also financially viable if only the recovery of cost for operation and maintenance is considered.

The alignments subject to the approximative technical analysis in this study serve as placeholders for thinkable solutions to connect Almaty and Issyk-Kul. Now that the potential for economic development has been shown, a feasibility study should be the next step to narrow down the possible alignments and more thoroughly determine their cost. It should thereby not be limited to the alignment chosen for the EIA.

Once the capital as well as operation and maintenance costs are more precisely determined, a financial analysis should look into options to recover cost and show financial viability.

Using the MRIO economic model for the ABEC region developed under this study, economic viability should again be examined on the basis of cost estimates.

1

Introduction

1.1   Almaty–Bishkek Economic Corridor

1.   The Almaty–Bishkek Economic Corridor (ABEC) is the pilot economic corridor under the Central Asia Regional Economic Cooperation Program. The ABEC initiative taps into the economic potential of Almaty (Kazakhstan) and Bishkek (Kyrgyz Republic), which are only about 240 kilometers apart from each other.

2.   The ABEC pursues development mainly in three sectors: tourism, agribusiness, and connectivity. This economic impact assessment (EIA) for an alternative road between Almaty and Issyk-Kul aims at increasing connectivity between parts of the ABEC region, for which tourism is of special interest.

3.   The ABEC region (Map 1) comprises the two cities of Almaty and Bishkek, Almaty and Zhambyl oblasts on the Kazakh side of the border, and Chuy and Issyk-Kul oblasts on the Kyrgyz side. Whenever this study uses statistical data, the data will include a total of six geographic units: one city and two oblasts on each side of the border.

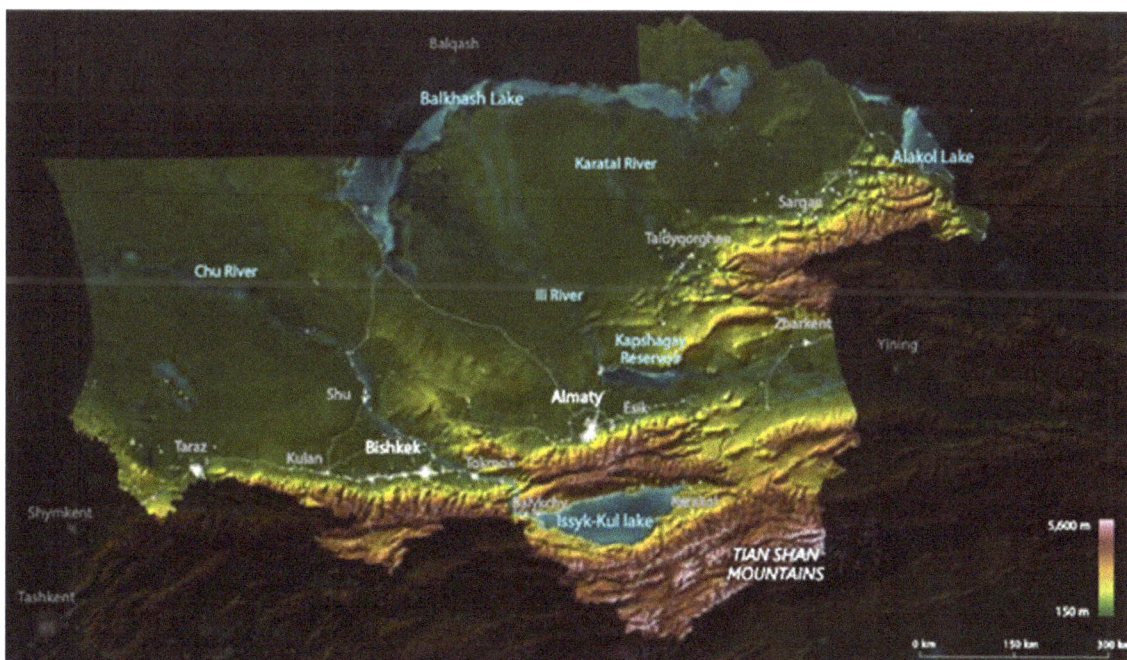

Map 1: The Almaty–Bishkek Economic Corridor: Geographic Definition

Source: DIVA-GIS. http://www.diva-gis.org.

1.2 Almaty and Issyk-Kul: Destination Details

4.   Almaty is a modern and vibrant city with a population of 1.9 million and has an international airport, which is Central Asia's busiest air hub. It has the potential to become increasingly attractive for tourism other than the established business travel, as it is situated in immediate proximity of scenic mountains and natural and cultural monuments. A considerable share of Almaty's population is able to afford traveling, making it a large and nearby potential market for visitors to Issyk-Kul.[1]

5.   Issyk-Kul is situated south of the Tian Shan mountain range and is famous for its large lake (fed by warm springs) and the impressive mountain ranges on both the north and south shores of the lake. Tourism has been the most important economic factor in Issyk-Kul, with many tourism facilities going back to the former Soviet era. With about 23,000 accommodation beds, Issyk-Kul accounts for 78.7% of all domestic and international tourist arrivals of the Kyrgyz Republic.[2]

Cholpon-Ata on Lake Issyk-Kul. Beach resort at the lakeside.

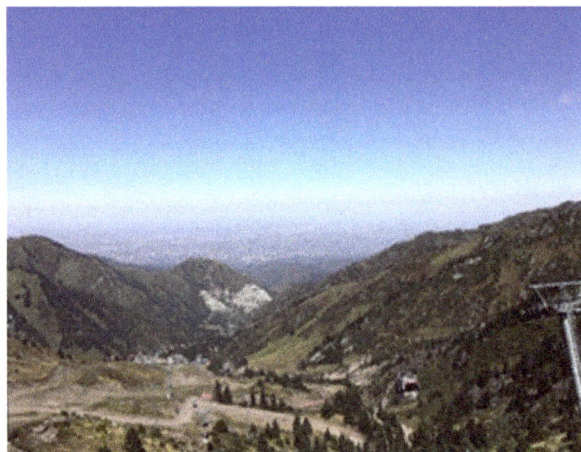

Almaty from Shymbulak Mountain Area. View into the valley toward Almaty.

6.   The influence the topography has on connectivity and economic bonds between Kazakhstan and the Kyrgyz Republic cannot be overstated. Lake Issyk-Kul is surrounded by mountain ranges with many peaks higher than 5,000 meters of altitude. Issyk-Kul can only be accessed from the east (via road) and from the west (via road and railroad).

7.   While the direct distance between Almaty and Cholpon-Ata on the north shore of Lake Issyk-Kul is only 80 kilometers, the actual travel distance is 460 kilometers, as the road bypasses the Tian Shan mountain range. The average travel time is 6.5 hours, not including border crossing time, reported to often be in the range of 1 hour. While the most direct route crosses the border near Kant (Kyrgyz Republic), that border crossing point (BCP) is largely used by trucks on their way through Naryn to the People's Republic of China and is therefore avoided by other vehicles. The Sartobe–Tokmok BCP used to be for bilateral exchange only but opened to third-country nationals in March 2020. The Korday–Ak-Jol BCP is often used by international travelers.

---

1   The average household income was 16% higher in Almaty in 2018 than in all of Kazakhstan.
2   Asian Development Bank (ADB). 2019. *ABEC Tourism Master Plan*. Manila.

8.   For travelers whose starting point or destination is the eastern part of Issyk-Kul, the existing road through Kegen (Kazakhstan) and Karkyra BCPs provide a more direct route to and from Almaty and may be more attractive. It is being rehabilitated and will be open year-round after the rehabilitation is complete.

9.   The only functional airport in the Issyk-Kul area is Issyk-Kul International Airport in Tamchy on the north shore of the lake, which has a few seasonal air connections to cities in Central Asia and the Russian Federation.

1.3   Alternative Road

10.   A new alternative road between Almaty and Issyk-Kul has been in the governments' and people's minds for decades; there were plans for a new road in the former Soviet era. In 2007, the European Bank for Reconstruction and Development (EBRD) developed a pre-feasibility study, which analyzed three alternatives in the area between Uzunagash (Kazakhstan)–Kemin (Kyrgyz Republic) and estimated their cost. However, the study was not followed by a more detailed feasibility study. In 2012, the governments of Kazakhstan and the Kyrgyz Republic agreed on a memorandum of understanding committing to build an alternative road between Almaty and Issyk-Kul.

11.   An alternative and more direct road connecting Almaty and Issyk-Kul would lower travel times and travel cost, inducing more frequent travel between the two destinations. This can be expected to have effects on the regional economy on both sides of the border. A new road is intended to be a tool for regional economic development, as it may lead to additional spending, especially in tourism, and stimulate investment. All of these activities would contribute to economic growth.

12.   The EIA is based on three alignments of an alternative road crossing the Tian Shan mountain range (Map 2). While it must be based on specific alignments and their estimated cost, all considered alignments should be understood as representing possible options, which may be planned and complemented in more detail in future studies. The economic impacts of any considered alignment are at the center of this study's interest. It will inform the governments of the Republic of Kazakhstan and the Kyrgyz Republic as well as private-sector stakeholders about the potential economic outcomes an alternative road would have.

13.   While the alignments, as presented in Map 2, all connect Almaty and Issyk-Kul, they are in fact situated far from each other and may serve different transportation markets. However, they would all serve visitors originating from the Almaty region or Almaty International Airport and traveling to Issyk-Kul or vice versa. The different alignments are presented and analyzed in more detail in Chapter 3.

Map 2: Existing Almaty–Issyk-Kul Road and Three Alternatives

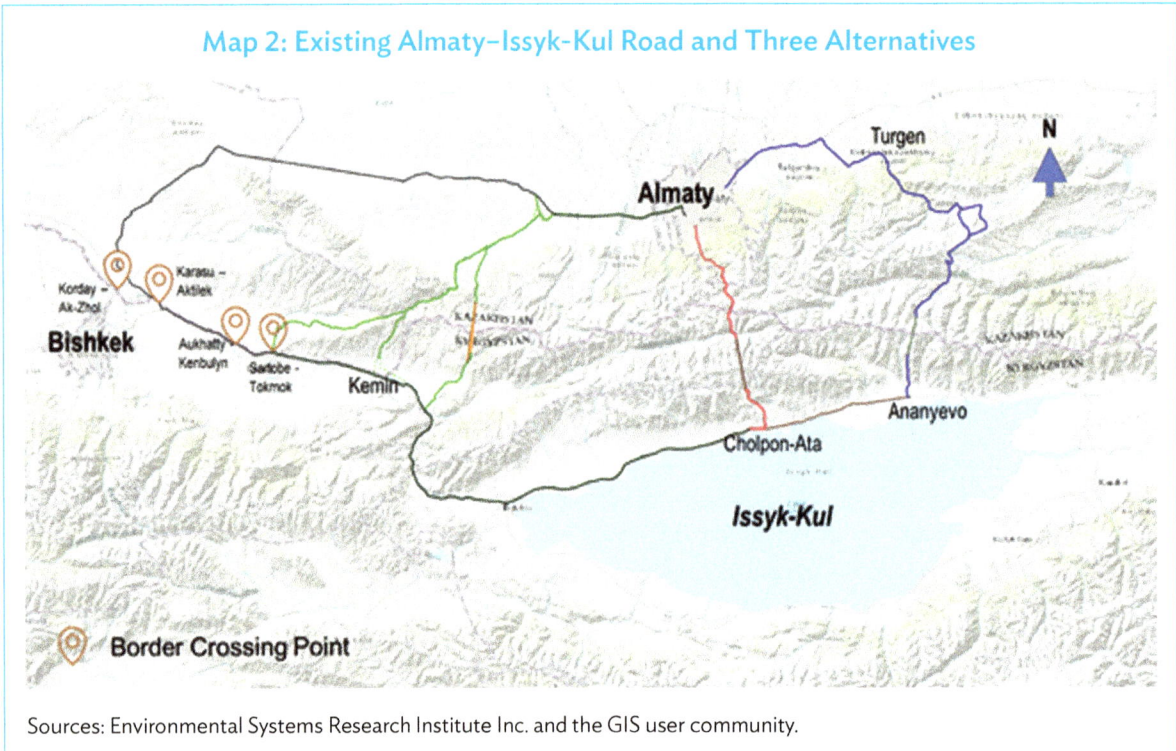

Sources: Environmental Systems Research Institute Inc. and the GIS user community.

2

Objective and Study Approach

2.1 Objective

14.   The EIA will enable ADB and the governments of Kazakhstan and the Kyrgyz Republic to assess the economic outcomes of an investment in a new alternative road between Almaty and Issyk-Kul. The results of the study should demonstrate the relationship between the investment, the traveler benefits, and the short- and long-term economic impacts on each side of the Kazakh–Kyrgyz border. The EIA is also a tool for government and stakeholder involvement, and it can be further used to communicate options concerning economic development and financing.

2.2 Study Approach

15.   This report documents the critical steps, which precede the actual EIA, as well as the methodology and results of the EIA. Data collection as the foundational basis of the study is described in Chapter 2, section 2.3. A comprehensive, albeit approximate, analysis of the different alignments and their cost follow in Chapter 3. Based on travel times resulting from the selected alignments and on additional considerations, changes in travel behavior are assumed and travel demand is estimated (Chapter 4).

16.   The EIA, described in further detail in Chapter 5, is designed to integrate all relevant impacts on the regional economies in both countries. The impacts are summarized and presented in the form of the return on investment (Chapter 6). Financing options for the alternative road are laid out in Chapter 7.

2.3 Sources

17.   The main data sources for this study were government agencies of Kazakhstan[3] and the Kyrgyz Republic.[4] Each agency endeavored to respond to the data requests, but obviously not all of the requested data was available, leaving data gaps. Publicly available statistics were retrieved, and data related to specific road infrastructure projects were received from various sources.

18.   The data obtained included demographic data for cities and oblasts, economic data about employment and income, tourism data (using the ABEC Tourism Master Plan as an additional source), infrastructure data, traffic data (e.g., for current or recent road projects), and financial and tax data. Legal information was also

---

3   Committee of Tourism Industry of the Ministry of Tourism and Sport of the Republic of Kazakhstan, Roads Committee of the Ministry of Industry and Infrastructural Development, and KazAutoZhol Joint Stock Company.
4   Department of Tourism of the Ministry of Culture, Information of Tourism of the Kyrgyz Republic, Ministry of Economy, Ministry of Transport and Roads, and National Statistics Committee of the Kyrgyz Republic.

useful. The EIA is based on input–output economic tables, which were used to build a multiregional input–output (MRIO) economic model to determine magnitudes of economic impacts caused by the alternative road.

19.   Meetings were held with representatives of the involved government agencies of both countries. During the initial phase in August 2019, meetings held in Bishkek and Nur-Sultan aimed at presenting this study's scope and learning about the respective agencies' perspectives on an alternative road between Almaty and Issyk-Kul. A joint meeting, held during project implementation in Almaty in November 2019, produced answers to specific questions regarding design of the alternative road and travel demand. Toward the end of the project, meetings with representatives of agencies provided feedback on the EIA's findings and a draft of this report.

20.   Additional meetings were held with individual nongovernment organizations during the initial phase of this study to obtain information about current conditions and the organizations' perspectives on any such road.[5]

21.   Interviews were conducted in the initial phase of the project, especially with individuals with firsthand knowledge about the tourism industry in Issyk-Kul, to learn about current conditions in the regional industry. Additional interviews with individuals in Almaty, Issyk-Kul, and Bishkek provided information on their travel behavior and how they spend their leisure time (Chapter 4, section 4.3.2).

---

5   Union of International Road Transport Carriers of the Republic of Kazakhstan, Kazakhstan Auto Transport Union, and Kazakhstan Tourist Association.

3

Alignment Analysis and Cost Estimates

3.1 Overview

22.   The study includes three different alignments: the western, direct, and eastern alignments. The western alignment (Map 2 and Table 1) corresponds to the route analyzed in the 2007 EBRD study (para. 10). It is longer than the other alignments, as it tries to bypass the Tian Shan mountains and crosses them at a lower altitude than the other alignments. The direct alignment is, as its name indicates, the most direct route, and it exists as a mountain path and has been in people's minds as a possible option for a road alignment for a long time. It includes two mountain passes at high altitudes. Considering the eastern alignment coincides with plans to increase access to the Turgen skiing resort in Kazakhstan, which is planned to be developed in Kazakhstan.

Table 1: Three Alternative Road Alignments

Alignment	Road	Length[a]	Comment
	Almaty–Bishkek–Cholpon-Ata	460 kilometers	Existing motorway via Bishkek
	Almaty–Kegen–Cholpon-Ata	470 kilometers	Existing route via Kegen
West	Almaty–Uzynagash–Kemin–Cholpon-Ata	260–351 kilometers	Sub-routes through New Kastek pass (2,461 meters) or Masanchi (maximum altitude 2,308 meters)
Direct	Almaty–Ozernyi pass– Chon-Kemin–Kungei Alatau tunnel–Baktuu-Dolonotu–Cholpon-Ata[b]	85 kilometers	Direct alternative via Ozernyi pass (3,514 meters), crossing Chon-Kemin National Park; tunnel at 3,000 meters through Kungei Alatau
East	Almaty–Turgen–Assy Plateau–Shilik River valley–tunnel to Baysar Valley–Ananyevo–Cholpon-Ata	216 kilometers	Synergies with Turgen ski resort, crossing Sarytau pass, crossing Kolsaiskie National Park, tunnel at 2,900 meters through Kungei Alatau

[a] Length of road connection between Almaty and Cholpon-Ata.
[b] The Ornok-Almaty route would be another option for the direct alignment, with possibly similar costs and economic impacts.
Sources: Google Maps and Google Earth.

23.   As this study is not trying to replace a feasibility study for the alternative road, it examines only potential approximate alignments for three distinct options. It is not possible or required to explore each of the alignments in great detail. Instead, the task of this analysis is to display alignments as reasonable assumptions of an alternative road between Almaty and Issyk-Kul as a basis for the EIA at the center of this study. More than one alternative has been developed for the western alignment to show a range of solutions with different technical characteristics, travel times, and costs.

3.2 Standards and Assumptions

24.   Development of the alignments and the cost estimates were made according to the Kazakh road classifications.[6] Road category III is mostly used for mountain roads. This implies a design traffic flow of 2,000–6,000 vehicles per day. The gradient along the longitudinal road profiles is generally kept below 12%. Steeper segments are accepted to limit technical challenges in mountainous terrain. Most of the alignments are designed to serve passenger cars, buses, and light trucks only, but no heavy trucks will be allowed for the purpose of keeping the road safe, fast, and comfortable for passenger transportation.

Mountain road construction. Workers are rehabilitating a road in the Kyrgyz Republic (photo by Asian Development Bank).

25.   While asphalt roads are more comfortable, allow for higher speeds, and save vehicle operating costs, they also require higher up-front investments than gravel roads. To explore the magnitudes of that difference, both gravel and asphalt solutions are included in the analysis of the western alignment. Gravel roads may either serve as an interim or as a permanent solution.

26.   Some of the alignments include tunnels, even though these complex infrastructure elements entail higher costs than road infrastructure in general. However, some of the considered alignments seem to be hardly feasible without tunnels that cap the most mountainous and technically challenging segments, including the presence of glaciers. Tunnels limit the maximum altitude to be surpassed by a mountain pass and thereby shorten trips and enable better road safety during seasons with snowfall.

27.   It is assumed that a tunnel, as part of the alignment, has to be safe and comfortable. Anything short of that would deter people from using the alternative road and limit economic impacts. This means that tunnel design has to include ventilation, lighting, and emergency exits.

---

6   Government of Kazakhstan. 2013. *Design Guidance of the Republic of Kazakhstan: SP RK 3.03-101-2013 "Highways and Roads."* Nur-Sultan.

28.   Roads at high altitudes in Central Asia are affected by low temperature, snow and ice. Snow and ice limit the use of any mountain road in the winter. At high altitudes, comprehensive snow and avalanche protection (e.g., tunnels, snow protection galleries, and avalanche barriers) and intensive snow cleaning would be required to keep mountain passes open in winter, entailing the respective higher capital investments and costs for operation and maintenance. However, since the desire for a year-round open connection between Almaty and Issyk-Kul is known and understandable, this study, based on snow and avalanche information for the Ala-Altai mountains, considers one of the alignments to be kept open year-round with a reasonable level of effort.[7]

29.   Each of the alignments crosses the border between Kazakhstan and the Kyrgyz Republic and requires a BCP. Future analysis will have to show if it is feasible to have only one BCP at the mountaintop or on one side of the border, or if it is necessary to have two separate BCPs on each side of the border with the top section of the mountain pass or a tunnel between them. More detailed information can be found in the Alignment Analysis (Supplementary Document that can be downloaded on http://www.almaty-bishkek.org).

3.3 Analysis of Each Alignment

3.3.1 Western Alignment

30.   The western alignment (Map 2), connecting Uzynagash in Kazakhstan with the Chuy Valley in the Kyrgyz Republic, was explored in the EBRD pre-feasibility study (para. 10). Partly based on that study, multiple alignment sub-routes were analyzed considering different technical parameters. To cover a wide variety of potential types, gravel and asphalt roads were considered, as well as solutions with and without a tunnel.

31.   Near Uzynagash, the alignment branches off from motorway A-2 coming from Almaty. Following existing roads that would need to be widened, the alignment to and beyond Kastek is the same for all sub-routes for more than 60 kilometers. From a point at an altitude of 2,308 meters, the route splits into two options.

32.   **Sub-route through New Kastek pass.** This option follows a tributary valley going south reaching the pass at 2,461 meters. Descending, the road follows a valley reaching Karasay Batyr village in the Chuy Valley, where a new BCP would be required to reach Kyrgyz territory. Almost all of this sub-route would be situated on the Kazakh side of the border. Costs are determined for both gravel and asphalt design.

33.   **Sub-route through Masanchi.** This option continues to the west from a point at 2,308 meters and descends to Keru village and further to Masanchi, from where an existing road connects Masanchi to the existing Sortobe–Tokmok BCP across the Chuy river. As the other sub-route, this road would also be almost entirely situated in Kazakh territory. Costs are determined for both gravel and asphalt design.

34.   Both sub-routes are suitable for passenger cars, buses, and light trucks only. Reaching altitudes above 2,000 meters, neither of the roads could be kept open year-round. It is assumed that both sub-routes could be kept open for about 6 months during June–November.

35.   The tunnel option for the western alignment would branch off from the road between Uzynagash and Kastek in Karakastek and ascend to an altitude of 1,600 meters, where a 14.6-kilometer tunnel would be

---

7   Out of 26 mountain passes in the Swiss Alps, 18 remain closed in winter. Only those at lower altitudes (up to about 2,000 meters above sea level) or those with no alternative route are kept open because of comprehensive snow and avalanche protection.

necessary to reach the Kyrgyz Republic. On the Kyrgyz side of the tunnel, the descending road would connect to the existing road between Ak-Tyuz and Kichi-Kemin and reaches the Chuy river (Map 3).

36.   This tunnel option is expected to be suitable for all vehicles and could be kept open year-round.

3.3.2 Direct Alignment

37.   The direct alignment (Map 4) would be laid out in the most direct way following one of the existing mountain trails between Almaty and Issyk-Kul, passing Big Almaty Lake. Two mountain ridges represent major obstacles for road construction: Ile-Alatau on the border between Kazakhstan and the Kyrgyz Republic, and Kungei Alatau on the Kyrgyz side. Crossing such difficult terrain would require a significant engineering effort. The construction of a tunnel through the Kungei Alatau mountain ridge is taken into consideration in this analysis.

38.   From Big Almaty Lake, the route follows the existing trail along the eastern shore. A closer look will be needed to explore the best option for a safe alignment in this segment. Ascending to Ozernyi pass, the road would reach the border between Kazakhstan and the Kyrgyz Republic at 3,514 meters. It is assumed that the BCPs for each country will be located at lower altitudes. The route descends along the Prokhodnoye gorge toward the Chon Kemin valley. A 16-kilometer-long tunnel at an altitude of roughly 3,000 meters would connect the Dolon-Ata and Koshko-Suu valleys. Descending to Lake Issyk-Kul, the road would connect to motorway A-363 in Baktuu-Dolontuu.

39.   The direct alignment would provide the shortest distance and travel time between Almaty and Issyk-Kul. However, with a mountain pass crossing 3,514 meters and a 16-kilometer-long tunnel at 3,000 meters, its construction, operation, and maintenance would be technically challenging. More specific design studies would be required to determine the detailed alignment and to prove its feasibility. An additional challenge is to cross Ile-Alatau National Park without disproportionately affecting the natural habitat and landscape.

Map 3: Western Alignment, Sub-Route with Tunnel

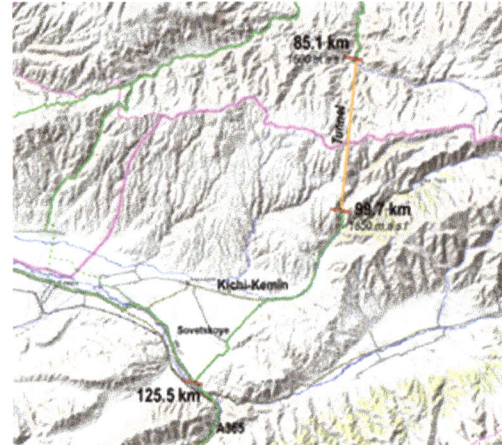

km = kilometer, masl = meter above sea level.
Source: OpenStreetMap.

Map 4: Direct Alignment, Southern Section

km = kilometer, masl = meter above sea level.
Source: OpenStreetMap.

40.   This alignment is suitable for passenger cars, buses, and light trucks only. Reaching altitudes of 3,500 meters, it could not be kept open year-round but only from June to November.

3.3.3 Eastern Alignment

41.   The eastern alignment (Map 5) would connect Turgen in Kazakhstan with Lake Issyk-Kul, providing potential synergies with access to a new ski resort planned near Turgen. Crossing the Ile-Alatau and the Kungei Alatau mountain ridges requires a tunnel through Kungei Alatau (as with the direct alignment).

42.   The alignment from Turgen follows the existing road to Batan village along the Turgen river and turns east toward the Assy observatory. The route ascends to the Sarytau mountain ridge and descends toward the Shilik river valley. This descent would require serpentines, galleries, and avalanche barriers. From the estuary of the Karasay river, the road would start ascending toward Kungei Alatau and reach the portal of a 12-kilometer-long tunnel at an altitude of 2,900 meters toward the valley of Orto-Baysar in the Kyrgyz Republic. The route continues to the Orto-Baysar

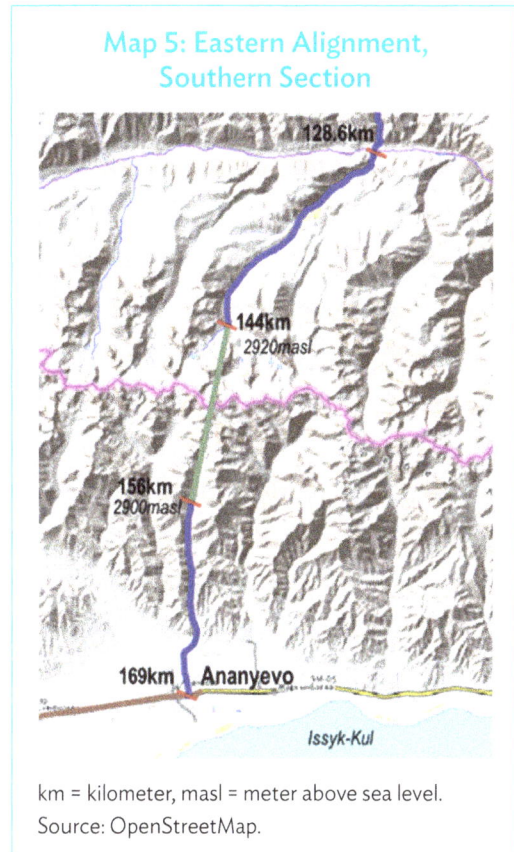

Map 5: Eastern Alignment, Southern Section

km = kilometer, masl = meter above sea level.
Source: OpenStreetMap.

river valley, which it follows toward Ananyevo on Lake Issyk-Kul, where it connects to motorway A-363. A more detailed analysis is required to determine a smoother descent.

43.   This alignment is suitable for passenger cars, buses, and light trucks only. Reaching an altitude of 2,900 meters, it could not be kept open year-round but only from June to November.

3.3.4 Travel Times and Characteristics of Alignments

44.   All alignments will considerably reduce travel times between Almaty and Issyk-Kul (Figure 1). Speeds along all road segments were estimated to determine average travel times between Almaty and Issyk-Kul (Cholpon-Ata). Different speeds for gravel and asphalt roads were also assumed.

45.   While the western alignment (comprising five sub-routes) lowers travel times by 20%–45% compared to the existing road through Korday, the direct alignment reduces travel times by almost 75% to approximately 100 minutes.

46.   Table 2 summarizes the characteristics of all considered alignments. While the western alignment sub-routes climb to altitudes up to 2,500 meters (much lower than the direct and the eastern alignments), they do not reduce travel distance and time to the same extent as the direct and eastern alignments. All three of the considered alignments include tunnels ranging from 12 to 16 kilometers. The western alignment sub-routes through New Kastek pass or Kastek do not include tunnels. Both gravel and asphalt options are considered for them to present a broader variety of solutions. While the western alignment option through Masanchi would make use of the existing Sortobe–Tokmok BCP, all other alignments would require construction of new

Figure 1: Travel Times by Alignment

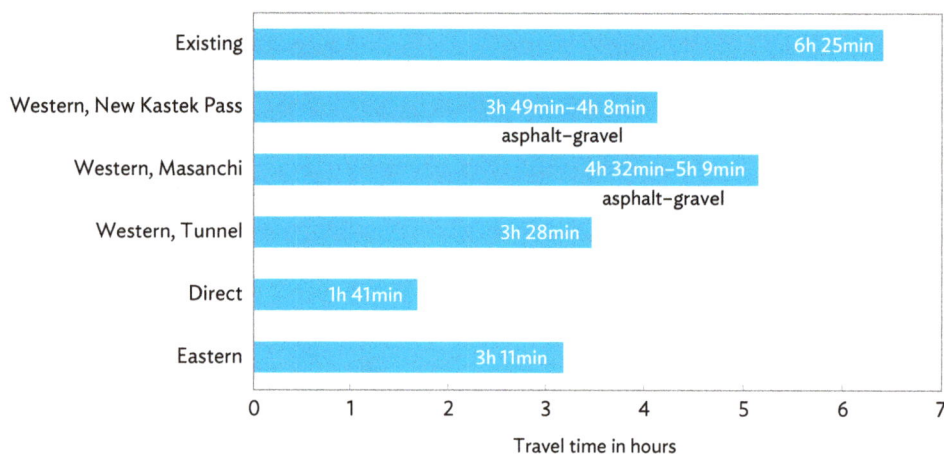

h = hour, min = minute.

Note: Excluding time spent at border crossing points.

Sources: Google Maps and consultant team analysis.

Table 2: Characteristics of the Alignments (Summary)

Category	West (Multiple Alternatives)			Direct	East
	New Kastek	Masanchi	Tunnel		
Length					
Reconstruction or new road	61 km	82 km	23 km	46.5 km	73 km
Tunnel			14.6 km	16 km	12 km
Highest elevation					
Northern ridge	2,461 m	2,300 m	1,850 m	3,514 m	2,970 m
Southern ridge				3,060 m	2,920 m
National border					
Altitude	850 m	830 m	2,610 m	4,355 m	4,155 m
BCP	1		1	1	1
National parks				Ile-Alatau	Kolsay Lakes
Almaty–Cholpon-Ata					
Distance	291 km	351 km	260 km	86 km	216 km
Travel time (without border crossing time)	Gravel: 4.1 h Asphalt: 3.8 h	Gravel: 5.2 h Asphalt: 4.5 h	3.5 h	1.7 h	3.2 h
Almaty–Karakol					
Distance	432 km	492 km	401 km	213 km	263 km
Travel time (without border crossing time)	Gravel: 6.2 h Asphalt: 5.9 h	Gravel: 7.2 h Asphalt: 6.6 h	5.5 h	3.4 h	3.6 h

BCP = border crossing point, h = hour, km = kilometer, m = meter.

Sources: Google Maps, Google Earth, and consultant team analysis.

BCPs. The direct alignment would cross Ile-Alatau National Park and the eastern alignment would cross Kolsay Lakes National Park.

3.4 Capital Cost Estimate

47.   The longitudinal profiles for each alignment (Figure 2 and Figure 3) are different: the western alignment alternatives cross altitudes in the range of 1,850–2,500 meters, the direct alignment reaches altitudes of up to 3,500 meters, and the eastern alignment reaches altitudes of 3,000 meters. Technical challenges, which are cost drivers, are also different.

48.   The level of effort to construct the road has been estimated for each segment individually. While some of the segments do not require any enhancement, other existing roads require widening or even reconstruction to provide the standards of a category II or III road. Road construction for new road segments is considered with the full range and volumes of work.

49.   Tunnel costs are included based on the analysis of tunnel projects in multiple countries. Per-kilometer tunnel costs vary largely, depending on specific circumstances and equipment, but there tends to be a higher per-kilometer cost for longer tunnels. Longer tunnels require more generous ventilation and safety equipment. Cost assumptions are based on tunnels in Asia as an approximation for safe and comfortable tunnels without using tunnel standards adopted in Western countries, which often include more sophisticated technical operations and safety features. The cost assumptions made for this study do not replace location-specific engineering and design considerations that would be part of a feasibility study. Only in a feasibility study would it be possible to assess costs and risks tied to the specific conditions of the terrain in a detailed way, and to describe the ways in which natural resources would be affected.

50.   For each BCP that would have to be installed, an additional cost of $2.5 million is added to the construction cost of the road.[8]

51.   Cost estimates for all alignments show a wide range (4), tunnels being the main driving factor of the differences. The sub-routes of the western alignment across New Kastek pass or Masanchi without tunnels are estimated to cause considerably lower costs than the other three alignment sub-routes with tunnels (Figure 4).

52.   These cost estimates should not be mistaken as the results of an in-depth feasibility study. For the purpose of this study, no thorough design of the alignments was desired. Rather, the cost estimates as presented reflect the approximate level of effort to construct an alternative road between Almaty and Issyk-Kul and will be considered as one type of spending leading to economic impacts (Chapter 5).

3.5 Operation and Maintenance Cost Estimates

53.   Besides capital investment costs for construction, operation and maintenance costs add to the total infrastructure cost. A statistical analysis of operation and maintenance costs in the Kyrgyz Republic shows an average operation and maintenance cost (including snow cleaning) of $4,366 per kilometers per year. Compared to the per-kilometer cost of recent road construction projects in the Kyrgyz Republic,[9] this corresponds to approximately 0.5% of construction cost. Internationally, an operation and maintenance effort corresponding

---

8   This cost assumption is based on plans for the new BCP in Karkyra.
9   Bishkek–Naryn–Torugart and Bishkek–Osh roads.

Figure 2: Longitudinal Profiles of Western Alignment Options

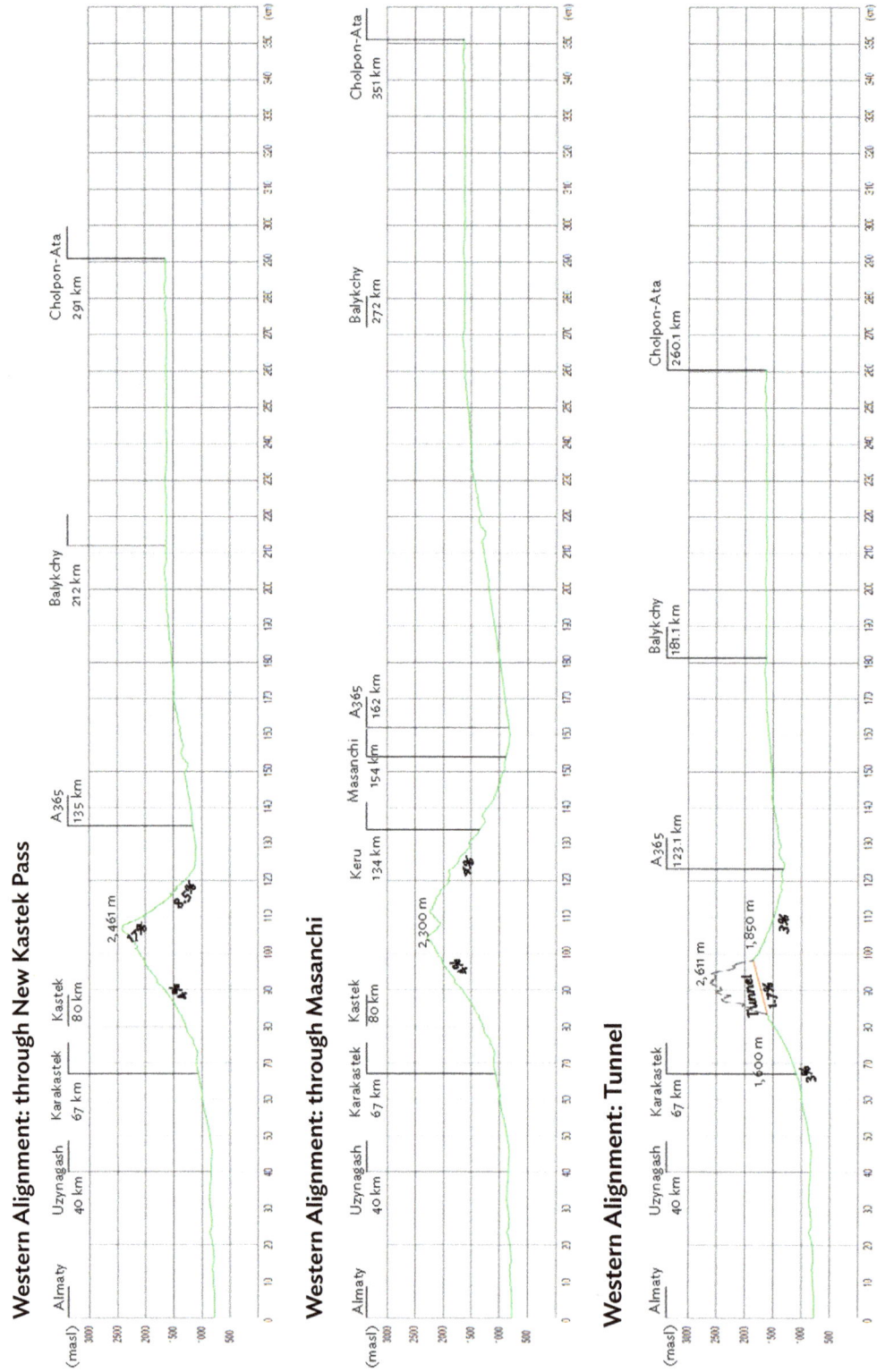

Western Alignment: through New Kastek Pass

Western Alignment: through Masanchi

Western Alignment: Tunnel

km = kilometer, masl = meters above sea level.

Source: Consultant team analysis.

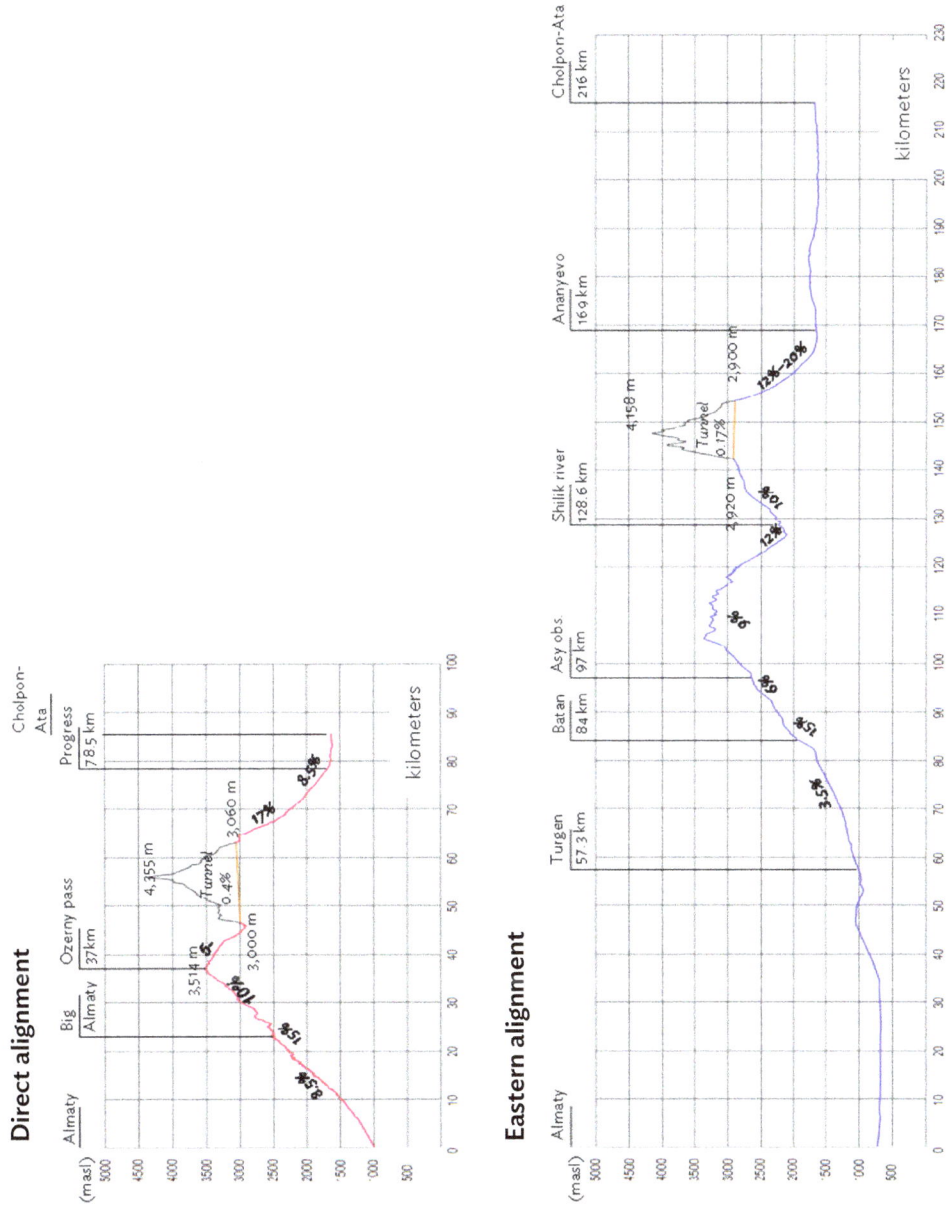

Figure 3: Longitudinal Profiles of Direct and Eastern Alignments

km = kilometer, masl = meter above sea level, m = meter, obs = observatory

Note: % signifies incline or decline

Source: Consultant team analysis.

Figure 4: Cost Estimates for All Alignments
($ million)

Alignment	Cost
Western, New Kastek Pass	81.2–93.8 gravel–asphalt
Western, Masanchi	92.5–112.8 gravel–asphalt
Western, Tunnel	517.7
Direct	586.6
Eastern	467.8

Capital Cost ($ million)

Sources: Google Maps and consultant team analysis.

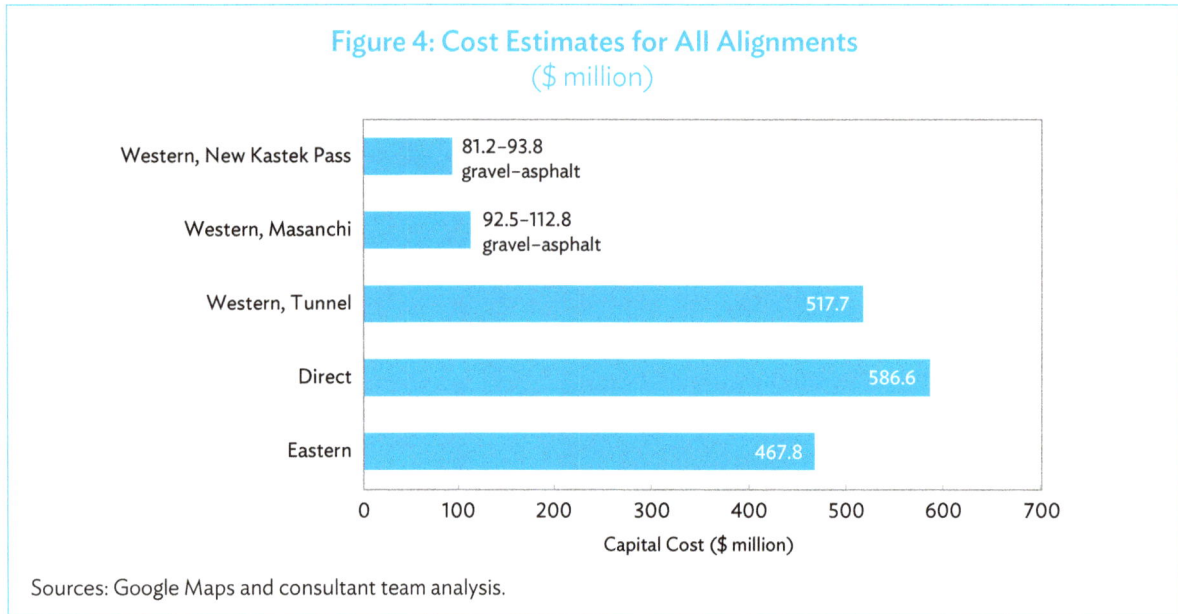

to 1%–2% of capital investment cost is considered sufficient to keep facilities in good long-term condition. Due to the mountainous terrain between Almaty and Issyk-Kul and to the high operation cost of tunnels, an operation and maintenance cost rate of 1.5% of total costs per year is used for this study. This means that for every $1 million spent on road construction, a further operation and maintenance cost of $15,000 per year must be expected.

3.6 Construction Period

54.   It is assumed that construction work could begin in 2025 for all alignments. Due to the different levels of effort to construct the road depending on the alignment, the estimated first year of operation falls in a range from 2029 to 2034. Table 3 shows what share of the construction cost is allocated on each country's territory.

Table 3: Estimated Construction Period and Allocated Cost of Construction by Alignment

Item	West					Direct	East
	New Kastek Pass		Masanchi				
	Gravel	Asphalt	Gravel	Asphalt	Tunnel	Direct	East
Start of construction	2025	2025	2025	2025	2025	2025	2025
End of construction[a]	2028	2028	2028	2028	2032	2033	2031
Start of operation	2029	2029	2029	2029	2033	2034	2032
Construction cost on Kazakh territory	90%	96%	90%	96%	55%	5%	55%
Construction cost on Kyrgyz territory	10%	4%	10%	4%	45%	95%	45%

[a] Based on the standard construction periods for road construction (categories II and III) provided in Government of Kazakhstan. 2013. Design Guidance of the Republic of Kazakhstan: SP RK 3.03-101-2013 "Highways and Roads." Nur-Sultan. Construction periods for tunnels assume a conservative average advance rate of conventional tunneling techniques (drill and blast) equal to 3 meters per day.

Sources: Consultant team assumptions and analysis.

4

Travel Demand Estimates

4.1 Sources for a Base Case Model

55.   No travel demand model was available to simulate the consequences of the new alternative road on travel behavior. Two models were considered, but, for various reasons, both turned out to be unusable. A simple desk model was built to include all available information about traffic volumes, growth rates, and origin–destination pairs.

56.   The following sources were used to feed a base case model:

(i)   **Pre-feasibility study Almaty–Issyk-Kul Road by EBRD in 2007.** Traffic counts and origin–destination surveys on the roads to Balykchy and to Naryn (Kyrgyz Republic), west and southwest of Issyk-Kul. The pre-feasibility study was not published.

(ii)   **Border crossing point counts.** 2018 traffic counts for the most relevant BCPs of Korday–Ak-Jol close to Bishkek, and Karkyra between Kegen and Tyup–Karakol in the east of Issyk-Kul.

(iii) **World Bank appraisal document for the third phase of the Central Asia Regional Links Program.**[10] Data point for traffic volumes and winter closures at this Karkyra BCP between Kegen and Tyup.

57.   The consultant team conducted its own traffic counts to complement and validate the data points found in studies mentioned in paragraph 56, by providing a cross-reference. Traffic was counted in 10 different locations in November 2019, providing two separate 1-hour samples per location. The results were adjusted for the time of day and season.[11] These results were mainly used to verify older or singular traffic count information.

58.   The Almaty–Bishkek Economic Corridor (ABEC) Tourism Master Plan represents a crucial source of tourism forecasts in the region (footnote 4). The master plan developed five scenarios, some of which envisaged improved connectivity between Almaty and Issyk-Kul (Box). Scenarios II and III from the master plan are used as an indication for the magnitude of tourism growth in the region, partly enabled by an alternative road between Almaty and Issyk-Kul.

---

10   World Bank. 2018. Project Appraisal Document on *A Proposed Credit In The Amount of SDR19.70 million ($27.50 million equivalent) and a Proposed Grant in the Amount of SDR19.70 million ($27.50 million equivalent)* to the Kyrgyz Republic for the Third Phase of the Central Asia Regional Links Program. Washington, DC.

11   The road segments included in the traffic counts were as follows: Almaty-Korday, Kemin-Balykchy, Balykchy-Cholpan-Ata, Cholpan-Ata-Ananyevo, Ananyevo-Korday, Korday-Almaty, Uzynagash-Kaynazar, Almaty-Esik (Kuldzhinsky tract), Almaty-Turgen (Kuljin tract), and Turgen-Kaynazar (Talgar tract).

Box: Summary of the Two Scenarios from the Almaty–Bishkek Economic Corridor Tourism Master Plan

Scenario II presupposes implementation of all policy-contingent and non-reform-dependent measures considered in the Tourism Master Plan. Non-reform-dependent measures include urban improvements, Silk Road signage, or support for Manas Airport (Bishkek). Policy-contingent measures are, besides the Almaty–Issyk-Kul alternative road, improved border crossing procedures and better conditions for Almaty Airport to develop its function as a hub for Central Asia. Under this scenario, the number of tourist arrivals in the Almaty–Bishkek Economic Corridor region would increase to 8.4 million in 2030 (a 10% increase compared to the base case) and 10 million in 2040 (a 13% increase). Foreign tourist arrivals and spending increase faster in Kazakhstan than in the Kyrgyz Republic.

Scenario III comprises implementation of all policy-contingent measures and all priority projects. In addition to the measures included in scenario II, improvements to tourism infrastructure and promotion as well as improved legal frameworks and law enforcement are considered. This is the scenario that is expected to generate the most tourist arrivals in the Almaty–Bishkek Economic Corridor region: 14.9 million in 2030 (a 21% increase compared to the base case) and 23.3 million in 2040 (64% increase). Tourism growth is equally strong in both countries.

Source: ADB. 2019. *Almaty–Bishkek Economic Corridor Tourism Master Plan.* Manila.

4.2 Base Case 2030

4.2.1 Characteristics of the Base Case

59.   The base case describes a situation in 2030, up to which traffic volumes will continue to grow and in which projects currently under construction will be implemented but no new alternative road between Almaty and Issyk-Kul will exist.

60.   Traffic volumes are estimated based on the various data sources mentioned in section 4.1 and escalated using a compound annual growth rate of 3% for existing roads both along the Korday and the Karkyra routes. The growth rate is based on two studies (EBRD 2007, World Bank 2018 [footnote 12]) and proved to be consistent with the general magnitude of traffic volumes counted in November 2019 (para. 57).

61.   The only project with relevant impacts being implemented between 2020 and 2030 is the rehabilitation of the Karkyra road. It will increase the service quality of the road and allow for a higher average speed, attracting more travelers. Additionally, the new Karkyra BCP will be kept open year-round, which changes mobility options for travelers originating from or heading to the Karakol area east of Lake Issyk-Kul.

4.2.2 Traffic Volumes 2030

62.   Based on all the information mentioned in section 4.1, a base case is interpreted to show traffic volumes for the relevant existing roads between Almaty and Issyk-Kul along the Korday route to the west and the Karkyra route to the east. The considered network and cross-sections, for which sources in para. 56 or own traffic counts give information about traffic volumes, are displayed in Figure 5.

Figure 5: Considered Network and Segments with Traffic Volume Counts

Kennen — Targap — Uzynagash — Almaty

Korday

Bishkek — Tokmok

Kainazar   Talghar Esik Turgen

Karkyra

Kemin — Intersection — Balykchy — Cholpon-Ata — Ananyevo — Ayup

Naryn

Torugart

Lake Issyk-Kul

Karakol

Traffic count from study
Validating own counts

Sources: Consultant team analysis.

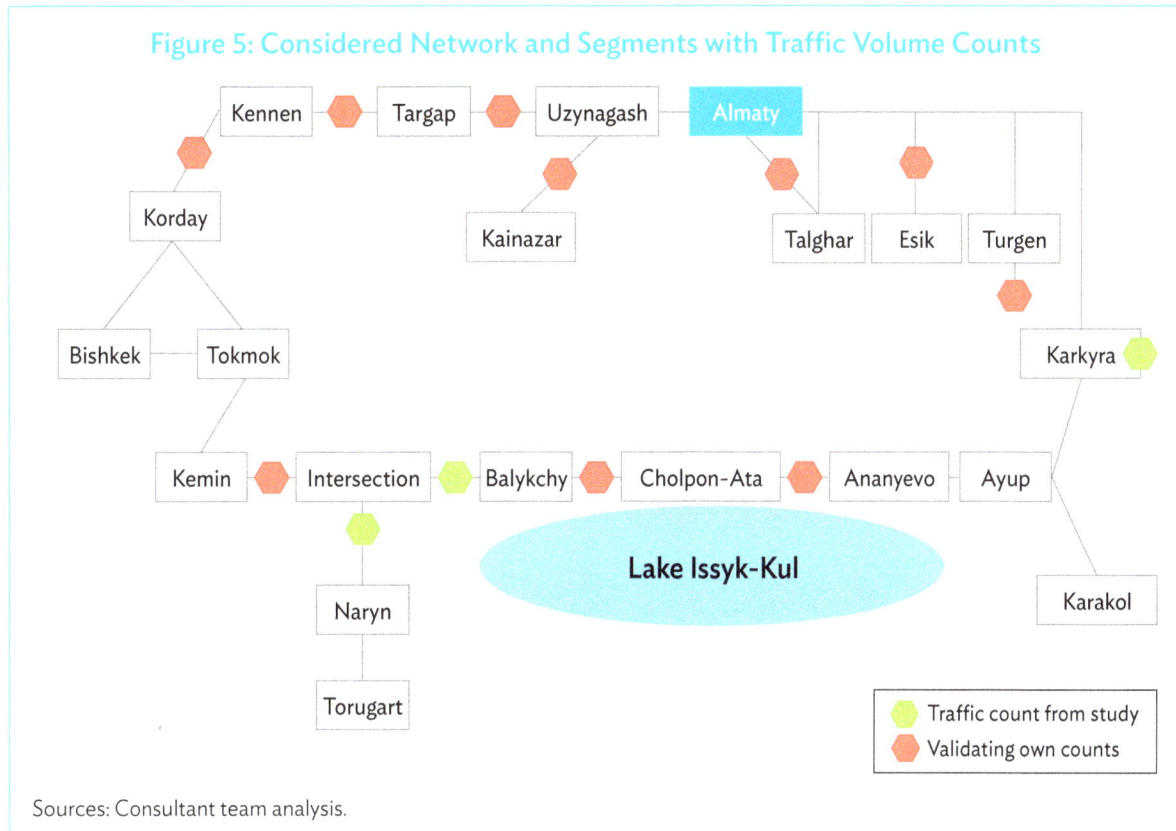

4.3 Changes in Travel Behavior

4.3.1 Introduction

63.   Changes in travel behavior depend on many factors, including rational, objectively traceable factors (like the attractiveness and accessibility of a destination or disposable income) as well as individual factors (like attitudes, the desire to travel, and personal impediments to do so). As with any personal behavior, travel behavior is difficult to predict.

64.   While changes in travel behavior are somewhat predictable when known transportation offers are improved (e.g., through expanding a highway to mitigate congestion), it is more difficult to predict travel behavior in cases of entirely new transportation offers that enable traveling in a way that was not possible before.

65.   This seems to also be true for the alternative road between Almaty and Issyk-Kul, especially since the road is expected to predominantly serve visitors to either the Almaty or to the Issyk-Kul region. For example, the current travel time of 6 hours and 25 minutes (without factoring in waiting time at the BCP) is too long for most people to consider the journey a weekend trip. The road would make weekend trips a new option for many potential travelers, although it is not known to what extent they will actually make use of it.

66.   Conducting a survey to find out about possible changes in travel behavior with the alternative road (and shorter travel times) was assessed not to be a viable way to source information. A stated preference survey could possibly show people's apparent desire to use such a new offer, but in reality, using the new offer

would be only one of many leisure options. Spending money on additional weekend trips between Almaty and Issyk-Kul, for example, would imply a willingness to pay for an option they are not currently aware of, from which a hypothetical bias could arise.[12] While stated preference surveys may be a good source of information in cases where people state their preferences about trips they undertake today, they do not necessarily reflect a scenario where assumptions need to be made to forecast new alternatives that are not currently available in the market.[13]

4.3.2 Approach

67.   A threefold approach was chosen to approximate future changes in travel behavior for the purpose of this study (paras. 97–99).

68.   **Part 1: Using principles known in transportation science.** As an alternative road would lower travel costs (out-of-pocket costs for fuel and vehicle maintenance as well as time costs), it would entice additional "induced" demand. Literature shows a variety of elasticity coefficients, depending on trip purpose, trip distance, and situation. Leisure or vacation trips are at the higher end, with elasticity coefficients ranging from –0.6 to –1.23.[14]

69.   **Part 2: Tying travel demand to activity forecasts (tourism).** The ABEC Tourism Master Plan provides forecasts of future tourism flows in five scenarios (Box). Scenarios II and III include the Almaty–Issyk-Kul alternative road among many other improvements. In both scenarios, flows of domestic but mainly foreign visitors to both the Kazakh and the Kyrgyz part of the ABEC region will increase. While scenario II is relatively conservative in its assumptions, scenario III leads to higher tourism forecasts than scenario II.

70.   **Part 3: Learning from interviews about individuals' travel behavior and experience.** Overall, about 30 interviews were conducted with individuals and tour operators in Almaty and Issyk-Kul as well as in Bishkek, which gave insights into travel behavior relating to trips to Almaty or Issyk-Kul. Some excerpts are displayed here:

> *"I usually go to Issyk-Kul for 10–12 days once a year using the services of tour operators. I prefer taking my summer vacation in Issyk-Kul rather than in Turkey, most importantly because of its climate—I enjoy the alpine climate of the lake.*
>
> *Currently, the road is very long and it takes a lot of effort to cross the border control point … If there were an alternative road with shorter travel time from Almaty to Issyk-Kul, I would be willing to go there almost every week with my own car. And I am more than certain that a lot of Almaty residents would do the same."*
>
> **—A. S., Almaty**

---

12   J. Loomis. 2011. What's to Know about Hypothetical Bias in Stated Preference Valuation Studies? *Journal of Economic Surveys.* 25 (2). pp. 363–370. and Hypothetical bias is a concept described as "the potential error induced by not confronting the individual with an actual situation." W. Schulze, R. d'Arge, and D. Brookshire. 1981. Valuing Environmental Commodities: Some Recent Experiments. *Land Economics.* 57. pp. 151–172.

13   E. Cherchi and D. Hensher. 2015. Workshop synthesis: Stated preference surveys and experimental design, an audit of the journey so far and future research perspectives. *Transportation Research Procedia.* 11. 154–164.

14   F. Dunkerley, C. Rohr, and A, Daly. 2014. *Road traffic demand elasticities: A rapid evidence assessment.* Cambridge, United Kingdom: Rand. For example, in Dargay, J. 2010. *The prospects for longer distance domestic coach, rail, air and car travel in Great Britain.* ITC Report. fuel cost-only elasticity coefficient of –0.79 for holiday trips longer than 150 miles was estimated. Elasticity of tourism demand with respect to the price of travel is close to unity (–0.98) in the long run according to a study about international tourism in Turkey. and A. Konovalova and E. Vidishcheva. 2013. Elasticity of Demand in Tourism and Hospitality. *European Journal of Economic Studies.* 4 (2). pp. 84–89. The average price elasticity for travel and accommodation costs in international tourism is about –0.6 to –0.8 to –2.

"If the time to travel from Almaty to Cholpon-Ata was reduced and the work of the border control point improved, I would definitely travel to Issyk-Kul more often, around 2–3 times during summer. I really like the nature around Issyk-Kul, so I would like to travel to this region during the weekends not only to enjoy the lake, but also for hiking in the mountains."

—A. A., Almaty

"The number of visitors would increase obviously, as my hotel has more Kazakh than Kyrgyz guests. And for people living in Issyk-Kul this would be a new opportunity to visit Almaty."

—Z. T., Issyk-Kul

4.3.3 Motivations and Impediments to More Frequent Travel Between Almaty and Issyk-Kul

71.   Because of time savings, travel from Almaty to Issyk-Kul or vice versa will become more attractive if a shorter alternative route is built. Due to the shorter travel times, weekend trips would be feasible for more people. It becomes clear from the interviews and from travel statistics how popular Issyk-Kul is as a nearby tourist destination. Predictions that improved connectivity will lead to a steep increase in the number of visitors to the region are supported by individuals stating their desire to travel more frequently between Almaty and Issyk-Kul.

72.   However, other destinations will remain attractive, or may become more accessible as well. Egypt, Turkey, or the United Arab Emirates, for example, are often named as preferred destinations for vacation trips of several days. Residents of Almaty and tour agents often mention Alakol as a favorite domestic destination. The travel times is 9 hours from Almaty to Alakol, but the road is being rehabilitated (Taldykorgan–Ust-Kamenogorsk), which may make trips more comfortable although not much faster.

73.   In comparison, the 1 million residents of Bishkek have different patterns of travel behavior. Travel times from the Kyrgyz capital to Issyk-Kul are approximately 2 hours and 30 minutes (Balykchy) or 3 hours and 30 minutes (Cholpon-Ata). Issyk-Kul is a preferred weekend destination for Bishkek residents, especially since various segments of the road have been rehabilitated or extended between 2000 and 2020. Interviews reveal that Issyk-Kul is a special place for some residents, who travel at least once per year, some of them several times in the summer and less frequently in winter for skiing. However, some interviewees find Issyk-Kul too far away for weekend trips. Other weekend destinations closer to Bishkek are mentioned, like Ala-Archa, Chunkurchak, Supara, and the ski resorts Kashka-Suu and Zil. Affordability, as well as the limited availability of different tourism product and experience offerings, may play a major role in the frequency and destination of weekend travel.

74.   The ABEC Tourism Master Plan and interviewees of this study state that there is a limited choice of weekend destinations from Almaty that provide basic tourist infrastructure. Tourist sites around Almaty (like Charyn Canyon or Assy) mostly lack that infrastructure today, though that still allows them to function as day trip destinations from Almaty. Travel times to some destinations are too long for short trips, as much for Issyk-Kul as for Kazakh destinations. Weekend trips of 3 days to Issyk-Kul are in low demand, according to travel agents. It can therefore be assumed that weekend travel to Issyk-Kul, enabled through the alternative road, would largely be a new option for Almaty residents who wish to travel during ordinary weekends.

75.   International tourists often combine two or more countries on a trip through Central Asia. According to analysis by tourism platform Indy Guide, more than 20% of travelers who visited the Kyrgyz Republic also went to Kazakhstan during the same trip.[15] Conversely, 14% of visitors to Kazakhstan indicated that they also visited the Kyrgyz Republic on the same trip. An alternative road would make it more likely for international

---

15   Indy Guide. Tourism Platform. http://www.indy-guide.com (accessed 2 March 2020).

travelers to combine destinations in and around Almaty with activities in Issyk-Kul, which would increase the attractiveness of the entire region on both sides of the border.

76.   While improved accessibility is an important factor for a destination's attractiveness, it is not the only one. There may be remaining impediments that make people decide not to travel to Issyk-Kul. According to interviews with hotel managers and travel agents in Issyk-Kul, professionalism and service quality industry may be among them. Kyrgyz law also provides only limited certainty for foreign property ownership, which may represent an obstacle for Kazakh citizens to invest in weekend houses in Issyk-Kul.

4.3.4 Conclusions Regarding Travel Behavior

77.   Using all available information to draw conclusions, the quantitative estimate of travel demand in Chapter 4, section 4.4 is based on the following:

(i)   An elasticity of –1.0 is used to estimate induced travel based on the literature reviewed (para. 68). All travel time and out-of-pocket savings will be spent on additional travel. While this elasticity may seem quite aggressive, it reflects values from studies about international tourism. It should also be considered that there is not currently a comparable destination to lake Issyk-Kul for potential weekend travelers from Almaty. Given Almaty's location and the ability of its middle class to travel, an excess demand is expected for weekend travel that cannot be satisfied.

(ii)   Tourist flow forecasts from the ABEC Tourism Master Plan's more conservative scenario II are matched by the alternative road in its most direct alignment. In scenario III, demand for travel between Almaty and Issyk-Kul is proportionally higher but is far from meeting the forecasts, as scenario III presupposes much more than an increase in connectivity.

(iii)   Weekend trips will become a considerable share of trips between Almaty and Issyk-Kul, with the assumption that 80% of additional trips could be weekend trips. The average duration of all trips would decrease accordingly. The alignments will each induce weekend trips to a different extent because of varying travel times; weekend trips on different alignments would therefore range from 29% to 68%. However, the consultant team refrains from defining a fixed time threshold for trips of shorter durations, because such a dichotomy is not supported by the literature (footnote 16).

(iv)   For all but one of the alignments, the alternative road can be kept open for up to 6 months (the western alignment option that includes the tunnel can be open year-round). For these alignments, the travel season would expand beyond the peak summer season (mid-June to mid-September), with about one-third of tourist travel demand happening outside of these 3 peak months.[16]

4.4 Demand Estimates for Alternative Road

4.4.1 Route Choice of Existing Travel

78.   Travelers between Almaty and Issyk-Kul largely use the shorter and more comfortable road through Korday, bypassing the mountain ridges to the west. A smaller portion of the traffic, however, uses the eastern Karkyra road.

79.   Based on an origin–destination survey conducted as part of the EBRD's 2007 pre-feasibility study, about 10% of travelers on the Korday route near Balykchy were known to be traveling between Almaty and the

---

16   Because there are no monthly tourism statistics available for Almaty or Issyk-Kul, this share is unknown. Based on interviews with tour operators and tourist facility owners, it is estimated that tourism is more concentrated on the peak season (mid-June to mid-September).

Issyk-Kul region. Applied to traffic volumes in the 2030 base case, an estimated 723 vehicles per day on an annual average daily traffic basis would choose the shorter alternative road instead of the existing road. This is true for all alignments, as they all would reduce travel time and travel cost and be more attractive than the existing road.

80.   Additionally, an estimated 519 vehicles per day on an annual average daily traffic basis would use the alternative road instead of the Karkyra road. This corresponds to the same number of vehicles switching from the Korday route after the Karkyra road is rehabilitated. With an alternative road, the same portion of travelers would again modify the route choice, if the alternative road follows either the direct or the eastern alignment. The western alignment, though, would not reduce travel times and cost for travelers from the Karkyra route. Figure 6 presents the volumes of diverted traffic by alignment.

Figure 6: Diverted Daily Traffic by Alignment

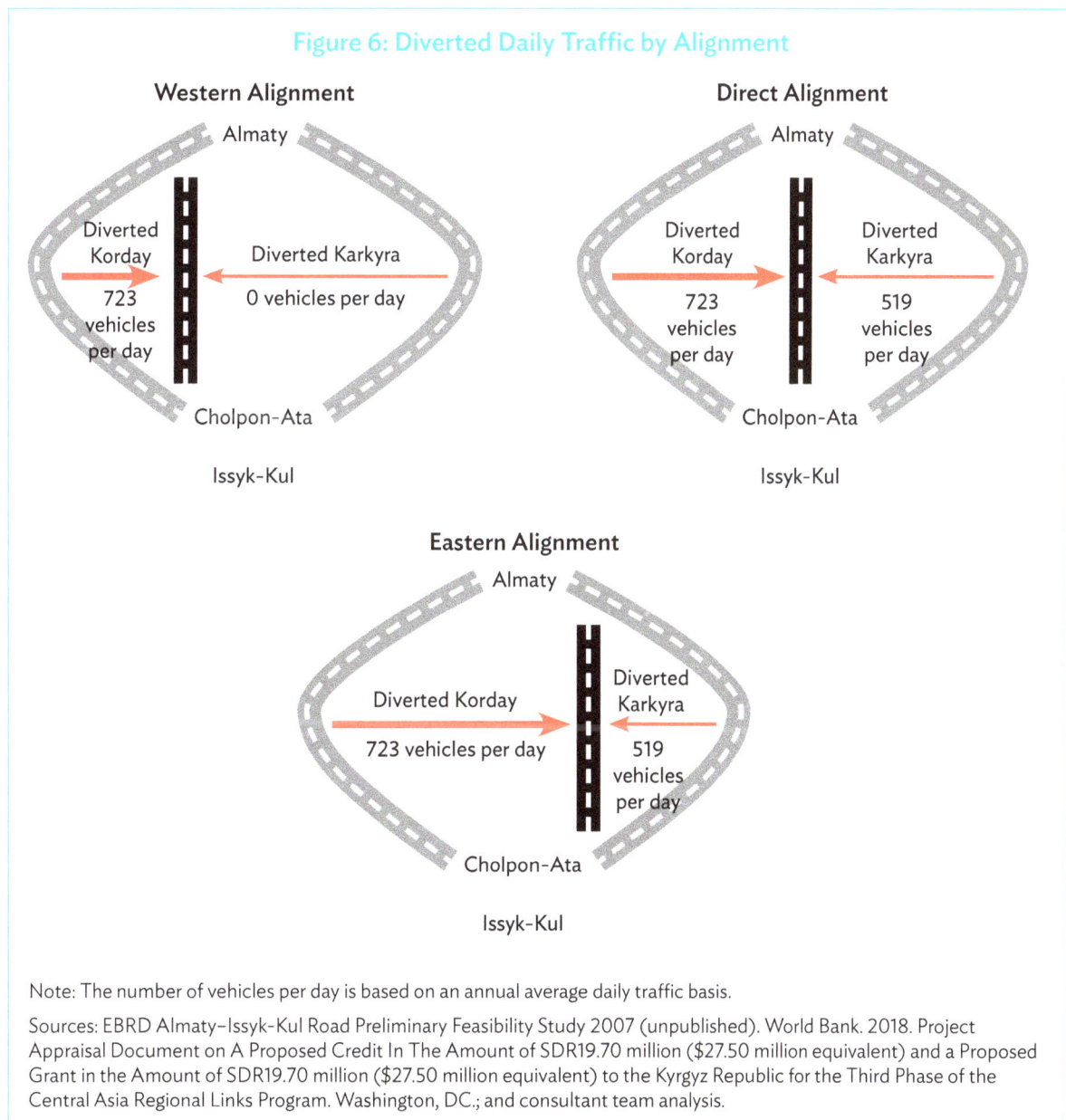

Western Alignment

Almaty

Diverted Korday — 723 vehicles per day

Diverted Karkyra — 0 vehicles per day

Cholpon-Ata

Issyk-Kul

Direct Alignment

Almaty

Diverted Korday — 723 vehicles per day

Diverted Karkyra — 519 vehicles per day

Cholpon-Ata

Issyk-Kul

Eastern Alignment

Almaty

Diverted Korday — 723 vehicles per day

Diverted Karkyra — 519 vehicles per day

Cholpon-Ata

Issyk-Kul

Note: The number of vehicles per day is based on an annual average daily traffic basis.

Sources: EBRD Almaty–Issyk-Kul Road Preliminary Feasibility Study 2007 (unpublished). World Bank. 2018. Project Appraisal Document on A Proposed Credit In The Amount of SDR19.70 million ($27.50 million equivalent) and a Proposed Grant in the Amount of SDR19.70 million ($27.50 million equivalent) to the Kyrgyz Republic for the Third Phase of the Central Asia Regional Links Program. Washington, DC.; and consultant team analysis.

4.4.2 Induced Travel

81.   Induced travel is additionally generated by reduced travel times and cost. More people will choose to travel to either destination, which will have become more accessible by means of the alternative road, or people who travel between Almaty and Issyk-Kul will decide to do so more often.

82.   The magnitude of induced travel depends on the reduction of travel times and cost.[17] It is therefore specific to each of the alignments presented in Chapter 3. Induced travel is inversely proportional to travel cost (para. 73). With lower travel costs, more travelers are expected to decide to travel between Almaty and Issyk-Kul.

4.4.3 Travel Generated from Economic Development

83.   The alternative road is expected to induce (i) additional trips between Almaty and Issyk-Kul and (ii) economic development in the areas benefiting from improved accessibility because of the alternative road. This economic development by itself generates additional trips with Issyk-Kul or the Almaty region as the destination as there will be more reason to travel due to improved or expanded activities at both ends of the alternative road. Additional tourists traveling between Almaty and Issyk-Kul will account for a large share of the newly generated travel, but other travelers will also benefit from the additional economic activity, e.g., workers, suppliers, and business travelers.

84.   Economic development as a result of the alternative road may consist of different components, which include the following:

   (i)   Additional demand for tourist services on both sides of the border because of lower travel cost may entice investments in capacity and quality of tourist infrastructure, leading to construction of new hotels, restaurants, and gas stations, and generating new jobs and income.
   (ii)   Traveler-oriented businesses may locate along the alternative road, leading to a more attractive route and generating trips made by workers and visitors.
   (iii) As a result of construction and supply of new or expanded businesses, these businesses and their workers as well as workers in other industries benefit from additional revenue and may invest in their own more productive businesses.

85.   Direct impacts from additional attraction of the Almaty and Issyk-Kul destinations, leading to indirect and induced impacts, are mainly expected in the tourism industry. Assumptions about travel generated by economic development are therefore largely based on the ABEC Tourism Master Plan and its visitor forecasts.

86.   The alternative road alone cannot exploit the entire potential presented in scenarios II and III of the ABEC Tourism Master Plan, as many other projects would have to be implemented to realize the full increase in visitors. A methodology was developed to estimate the share of the visitor flow increase because of the alternative road. It is assumed that the alignment offering the biggest travel time and cost savings (i.e., the direct alignment) could fully exploit the potential in scenario II. For all other alignments and for scenario III, all estimates of newly generated trips as a result of economic development are adjusted proportionally to match their impacts on travel times and vehicle operating costs.

---

17   In accordance with other studies and on the basis of own estimates, the following cost factors are applied: $5 per person-hour, $0.10 per vehicle-kilometer, average vehicle occupancy of 3.0 for passenger cars and 30.0 for buses.

4.4.4 Total Travel Demand

87.   In total, each alignment will carry diverted traffic from existing roads, induced travel, and newly generated travel from economic development. The numbers for all three effects are added up for all alignments and scenarios II and III from the ABEC Tourism Master Plan and are presented in Figure 7. Numbers are shown for a 6-month period, during which all alignments can minimally be kept open.

88.   For simplicity, numbers for 2030 are shown in Figure 7, even though not all alignments will be operational in 2030. Depending on the construction period, some alignments, especially those including tunnels, will be open to traffic only after 2030.

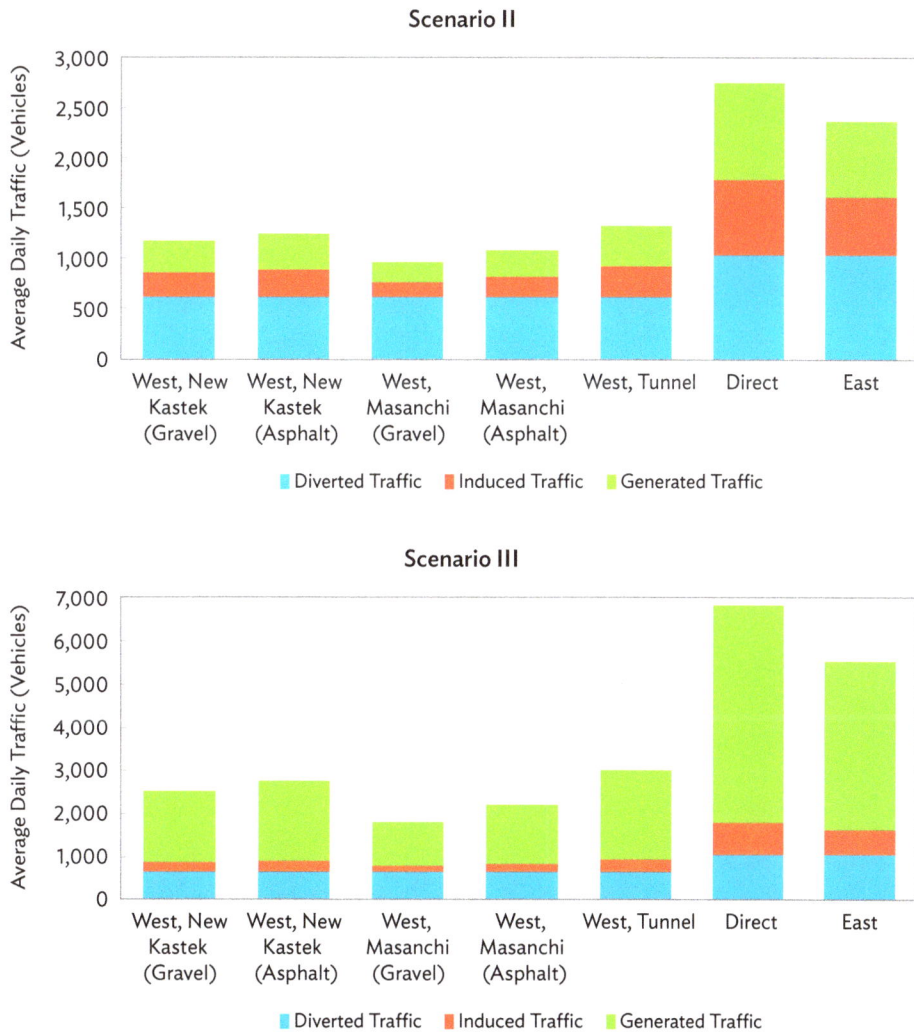

Figure 7: Estimated Traffic Volumes for Each Alignment and Policy Scenario (Average Daily Traffic for a 6-Month Period), 2030

Note: Policy scenarios refers to policy scenarios II and III from the Almaty–Bishkek Economic Corridor Tourism Master Plan.

Source: Consultant team analysis.

89.   As the alignments serve primarily tourism traffic, traffic volumes will be different between the summer peak and the off-season. Statistical charts for tourism routes are used as well as numbers of international tourists by month to combine an approximated distribution of travel demand across the year. Figure 8 expresses average daily traffic volumes across a 12-month period. The only alignment open in winter, the western alignment with tunnel, shows the smallest difference, as some (lower) travel demand can be expected in the winter months (Figure 10).

Figure 8: Estimated Traffic Volumes for Each Alignment and Policy Scenario (Average Annual Daily Traffic for a 12-Month Period), 2030

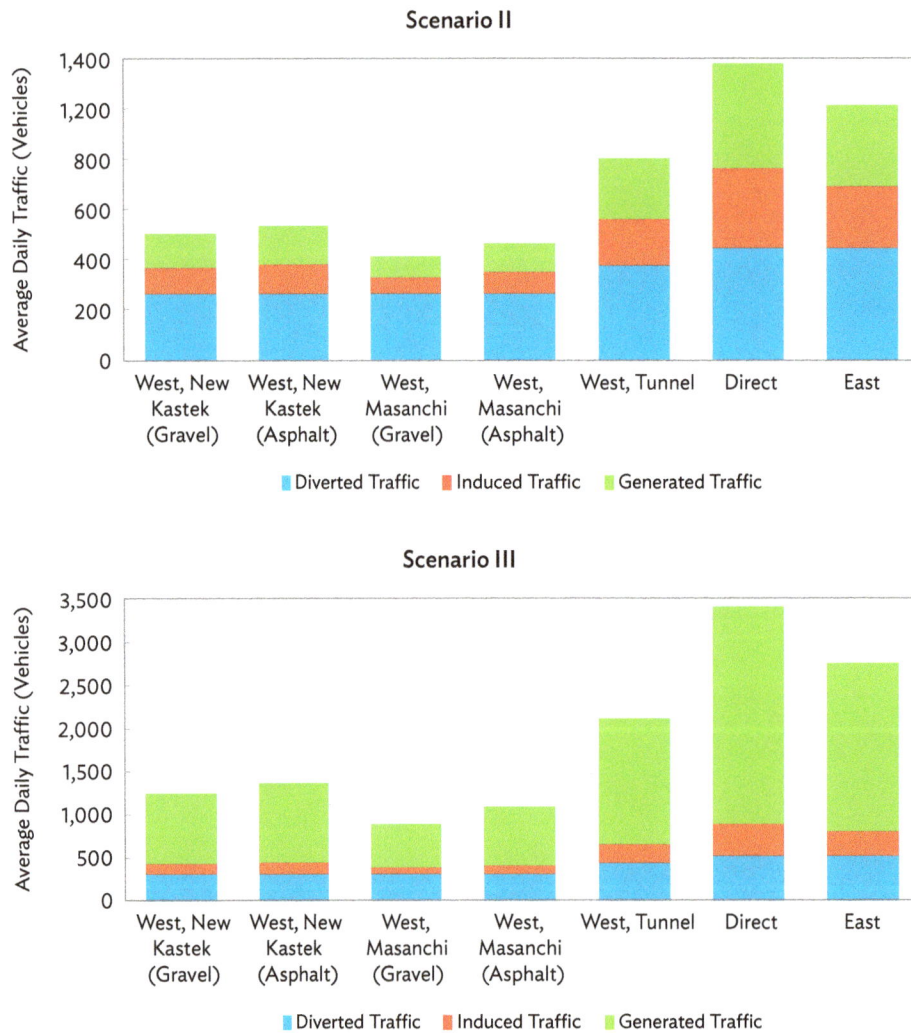

Note: Policy scenarios refers to policy scenarios II and III from the Almaty–Bishkek Economic Corridor Tourism Master Plan.

Source: Consultant team analysis.

90.   Figure 9 shows, for the direct alignment, a peak in August of nearly 4,000 vehicles per day (scenario II) and more than 9,000 vehicles per day (scenario III). This is the alignment with the highest traffic volumes, as it is the most direct route and therefore provides the biggest reduction in travel time and cost.

Figure 9: Traffic Volumes on the Direct Alignment in Scenarios II and III, 2030

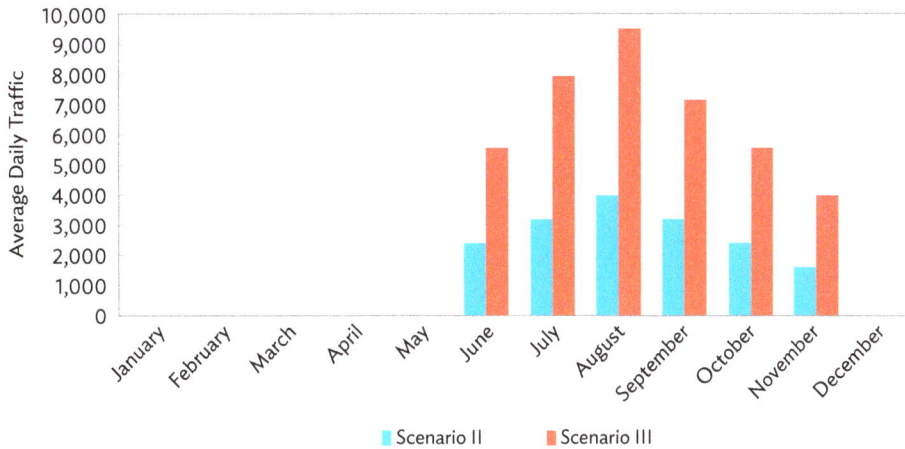

Note: Policy scenarios refers to policy scenarios II and III from the Almaty–Bishkek Economic Corridor Tourism Master Plan.
Source: Consultant team analysis.

91.   In comparison, the traffic volumes for the western alignment (with tunnel) are lower but expand over the entire year as this is the only alignment that would not require a planned winter closure (Figure 10).

Figure 10: Traffic Volumes on Western Alignment (with Tunnel) in Scenarios II and III, 2030

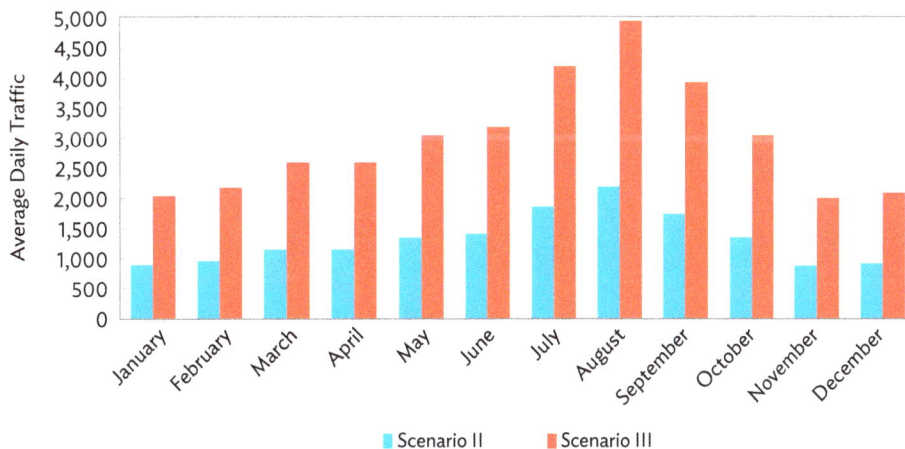

Note: Policy scenarios refers to policy scenarios II and III from the Almaty–Bishkek Economic Corridor Tourism Master Plan.
Source: Consultant team analysis.

92.   When assessing the level of resulting traffic volumes, comparisons to mountain passes in the Swiss Alps were made (Table 4). As an example, the Brünig Pass connects a catchment area in Central Switzerland with a population of about 1 million to a major tourism area, the Bernese Oberland, which includes major resorts like Grindelwald, Gstaad, and Interlaken. The Brünig Pass reaches an altitude of only about 1,000 meters and can be kept open year-round.

Table 4: Traffic Volumes on Swiss Mountain Passes (2017)

Mountain Pass	Road	Altitude (meters above sea level)	August ADT (vehicles/day)	AADT (vehicles/day)
Brünig	A 8	1,002	11,137	7,459
San Bernardino (Tunnel)	A 13	2,067	11,572	7,428
Grimsel	H 6	2,165	2,640	2,077
Flüela	H 28	2,383	3,428	2,125
Ofenpass	H 28	2,145	2,934	1,688
Julier	H 3	2,284	3,972	3,035
Furkapass	H 19	2,429	2,440	1,832
Oberalppass	H 19	2,044	2,475	1,502
Sustenpass	H 11	2,224	2,042	701
Nufenenpass		2,478	1,558	1,236
Klausenpass	H 17	1,948	1,438	1,080
Lucomagno		1,972	3,370	1,934
Col des Mosses		1,445	2,660	1,996
Bernina		2,328	4,358	2,053

AADT = average annual daily traffic, ADT = average daily traffic.
Source: Government of Switzerland, Federal Roads Office.

93.   Examples from similar situations in Central Asia or other neighboring countries could unfortunately not be found to compare the increase in visitor flows resulting from an improvement in accessibility. Less-developed tourist destinations (e.g., Svaneti, Georgia) showed much stronger visitor flows in the wake of an improved accessibility, but at a low level and were therefore not comparable to Issyk-Kul. In Western European countries, the easily accessible location close to strong tourism markets led to a large share of international mass tourism, which is not what can be expected in the case of Issyk-Kul. More detailed information can be found in the Travel Demand Estimates (Supplementary Document that can be downloaded on http://www.almaty-bishkek.org).

5 Economic Impact Assessment

5.1 Design

5.1.1 Overview

94.   This chapter describes the mechanics and methodology for calculating impacts on the economy. Figure 11 shows the core factors that drive the regional economic impacts: spending effects, traveler benefit effects, and economic development effects. It also shows the key factors and measures of those effects. Figure 12 is a flowchart showing the information sources and data elements used in calculating economic impacts.

95.   To arrive at a complete assessment of economic impact, analyses have been conducted covering three distinct but related economic views (paras. 96–98).

96.   **View 1: Spending effects (nonrecurring economic impacts from road construction.** The first view is the impact of construction spending for the road project. The construction costs are run through the multiregional input–output (MRIO) economic model (Chapter 5, section 5.1.2). Nonrecurring construction

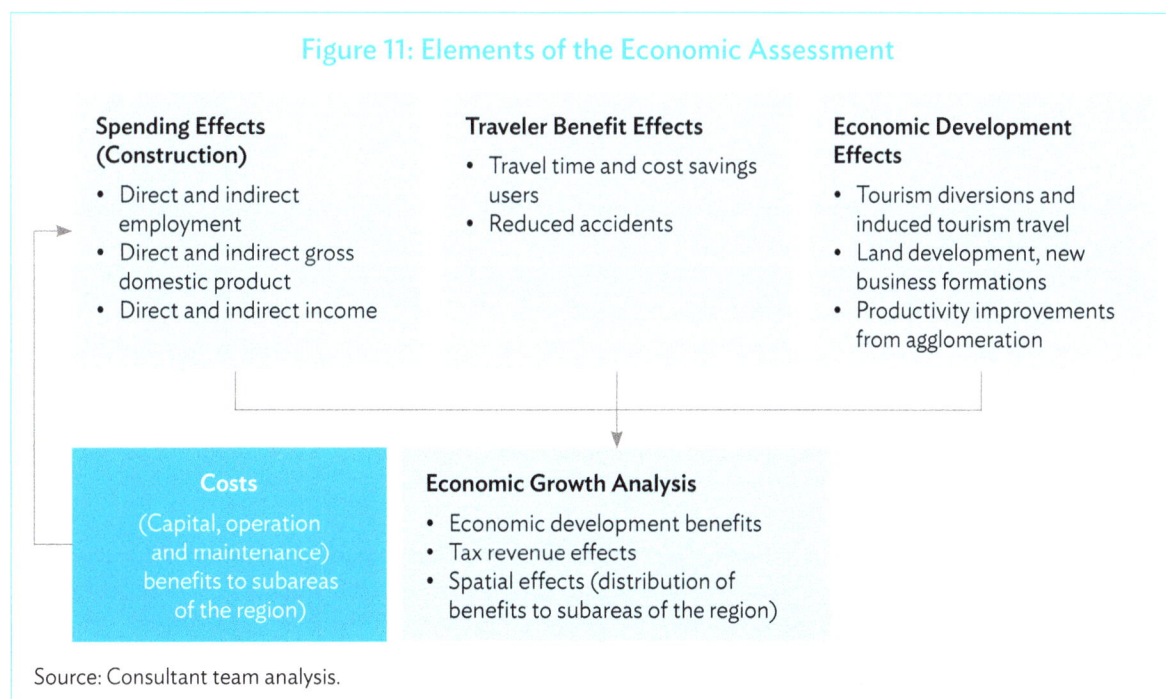

Figure 11: Elements of the Economic Assessment

Spending Effects (Construction)
- Direct and indirect employment
- Direct and indirect gross domestic product
- Direct and indirect income

Traveler Benefit Effects
- Travel time and cost savings users
- Reduced accidents

Economic Development Effects
- Tourism diversions and induced tourism travel
- Land development, new business formations
- Productivity improvements from agglomeration

Costs
(Capital, operation and maintenance) benefits to subareas of the region)

Economic Growth Analysis
- Economic development benefits
- Tax revenue effects
- Spatial effects (distribution of benefits to subareas of the region)

Source: Consultant team analysis.

Figure 12: Conceptual Overview of the Economic Impact Assessment Information Sources and Measures

| EIA for spending effects (construction) | Modeling and other impact analysis tools or resources | EIA for economic development effects |

Interviews

Increased frequency of existing travel | Diversion of tourism visits from other routes | Newly induced tourist travel

ABEC Tourism Master Plan

Capital Costs: Nonrecurring impacts during construction phase

Increased tourism and other visitor demand results in:
- increased land development (e.g., new hotels, second homes)
- increased visitor spending

National or Regional Input–Output Model

Direct, Indirect and Induced Economic Impacts
Employment, Personal and Business Earnings, GDP (Value-Added), Tax Revenue, Fiscal Impacts from Expanded Tax Base

Direct, Indirect and Induced Economic Impacts
Employment, Personal and Business Earnings, GDP (Value-Added), Tax Revenue, Fiscal Impacts from Expanded Tax Base

ABEC = Almaty–Bishkek Economic Corridor, EIA = economic impact assessment, GDP = gross domestic product.

Source: Consultant team analysis.

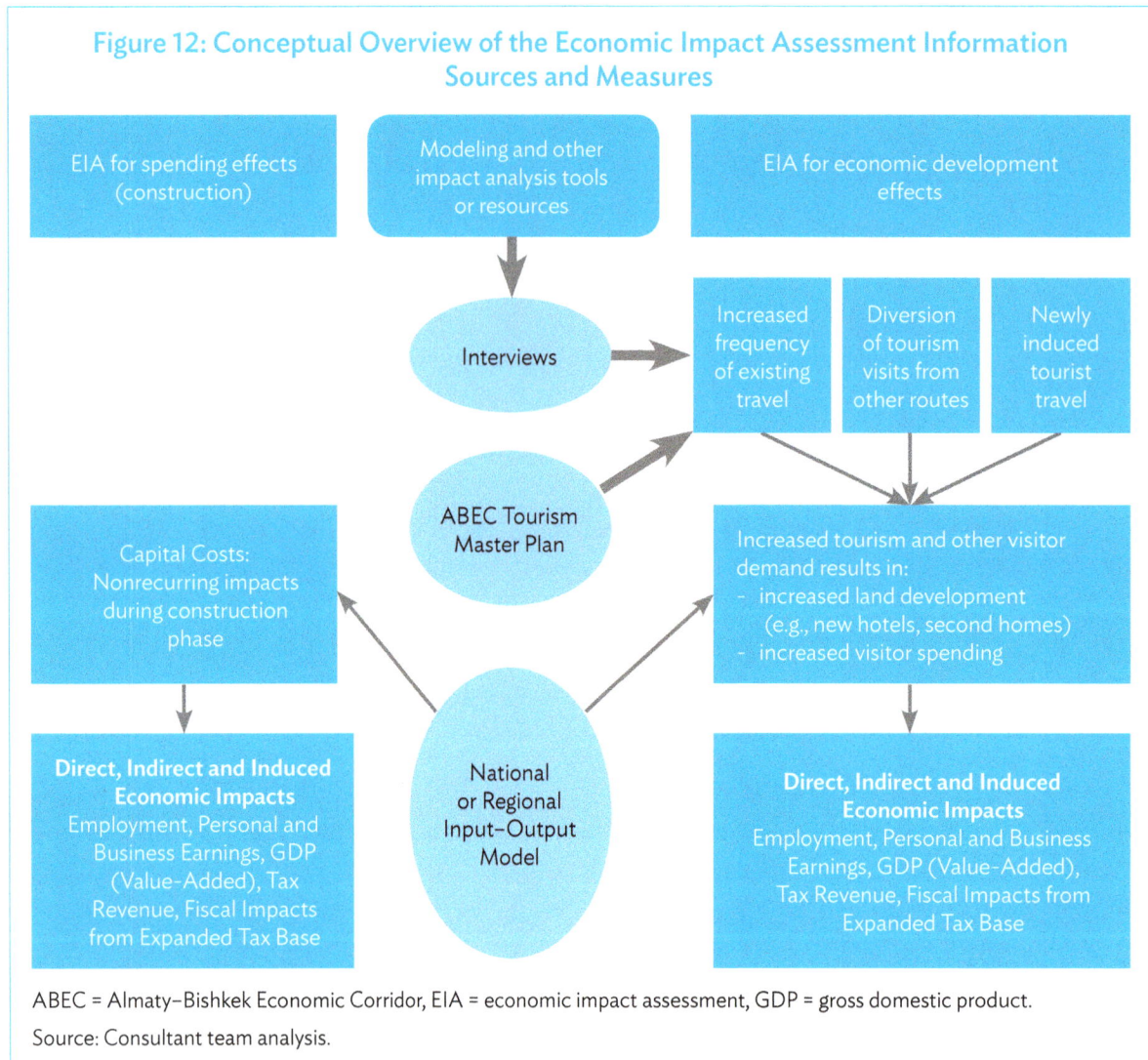

impacts also include construction of hotels and other tourism facilities arising from increased tourism demand, but these are included in view 3 (para. 98).

97.   **View 2: Traveler benefit impacts.** The second view takes a traveler benefit perspective, counting gains in terms of travel time savings and vehicle operating cost savings. The consultant team developed specific inputs such as the economic value of travel time and the specific vehicle operating costs based on national data, models such as the World Bank Highway Development and Management model, and research. This view does not count temporary effects of creating jobs related to road construction and operation, nor does it count effects of inward economic investment on local growth in the project area.

98.   **View 3: Economic growth impacts.** The third view involves the impact of enhanced accessibility on inward investment and hence growth of long-term economic activity within the study area. This view focuses on the impacts of enhanced tourism and tourism-related spending on critical measures of economic activity—measures typically used in national income and product accounts by most nations and regions. This third view may be regarded as the most important of the three, as it accounts for long-term impacts on the economy that

recur year after year, and that it provides a long-term and relatively permanent gain to the regional and national economies considered by this study. In effect, the third view provides what is typically referred to as an EIA, as it focuses on long-term job and household income benefits that are typically reflected by national income and product accounting in most countries.

5.1.2 Study Elements

99.   As suggested in Figure 12, the overall analysis is comprised of the following study elements:

(i)   **Develop travel demand forecasts.** As described in Chapter 4.

(ii)   **Compute traveler benefits.** These include benefits such as travel time and cost savings. Traveler benefits are estimated for a hypothetical year and interpolated for a period of years following construction. The cost savings lead to further spending in the economy, while the time savings enable broader impacts on travel demand and economic growth.

(iii)   **Develop a multiregional input–output model.** To adequately capture the complete impact of investments undertaken by both Kazakhstan and the Kyrgyz Republic, the consultant team constructed an MRIO economic model. The MRIO economic model explicitly factors in trade between the countries as a way of reflecting the economic effects of investment in one country on its neighbor. Like all other input–output models, it produces multipliers capable of portraying direct, indirect, and induced impacts, both within each country and across borders.

(iv)   **Address spatial issues and spatial impact disaggregation.** Within the MRIO economic model, the consultant team developed assumptions and methodologies for allocating the impacts to the Kyrgyz and the Kazakh part of the ABEC region. Allocation methodologies have been based on economic output measures for each region (output, value-added, income, and employment); anticipated cross-border flows; distance of the new road in each country for capital cost allocation; and available or developed traffic data, with impacts based on origin and destination information and assumptions.

(v)   **Develop alternative growth scenarios.** Based on refined and enhanced assumptions about the level and mix of increased tourism activity growth, scenarios aligned with the scenarios in the ABEC Tourism Master Plan were developed. These alternative scenarios include numbers of visitors, first-round spending associated with enhanced tourism (e.g., for hotels, restaurants, and other tourism services); and additional land development, focusing on new hotels and other accommodations, and additional tourism-oriented businesses such as restaurants, transportation, spas, and related activities.

(vi)   **Apply the multiregional input–output model to new construction and to additional direct tourism spending.** This will enable calculation of the full economic impacts over time, including outputs, value-added (closely related to enhanced GDP per country), personal income increases, and employment increases.

(vii)   **Develop fiscal (tax revenue) impacts for each country.** This will be based on information on the tax structure of each country and region.

(viii) **Conduct an enhanced economic return on investment analysis.** The primary focus of the study is an EIA of economic growth effects. Within that framework, an even more comprehensive return on investment (ROI) analysis is provided, which compares the full slate of economic benefits of each alternative against roadway investment costs. In this context, local economic growth is counted as a core element of the ROI, which is different from a traditional cost–benefit analysis (CBA) that focuses on transportation system efficiency. This form of ROI analysis provides the most robust possible perspective on the economic impacts and viability of the proposed road, which is appropriate given the focus of ADB on encouraging economic development through the more direct road between Almaty and Issyk-Kul (Chapter 6).

5.1.3 Informing Studies and Sources of Information

100.   Specific activities that inform the analysis include the following:

(i)   construction cost estimates of the alternative road for each alignment analyzed (Chapter 3, section 3.4);

(ii)   annual operation and maintenance cost estimates of each road alignment (Chapter 3, section 3.5);

(iii) construction of an MRIO economic model (developed by the consultant team based on assembled disaggregated economic data), with the basic processing to arrive at economic impacts involving estimation of first-order impacts, which represent direct final demand shocks;

(iv)   compilation of information on the taxation structures of both countries, such that fiscal effects of new development can be made based on changes in the relevant tax base for each tax; and

(v)   review of the ABEC Tourism Master Plan and conduct extensive interviews to refine the information obtained to develop estimates of projected expenditure impacts from existing tourism travel, and newly induced tourism travel.

101.   To adequately capture the complete impact of investments undertaken by both Kazakhstan and the Kyrgyz Republic, the consultant team constructed an MRIO economic model. An MRIO economic model is required because investment in one country will generate indirect effects on businesses in other countries, as supply chains spill over national borders. An MRIO economic model explicitly factors in trade between the countries as a way of linking inputs to production beyond any single nation's borders. Doing so allows modeling of how an increase in final demand in one country generates impacts on the economy of not just that country but on the economy of its neighboring country.

102.   Input–output tables served as the core data used to construct this model. These tables offered the unique advantage of having a common industry classification scheme for both countries, as well as explicitly estimating the demand for imported commodities at an industry level in the desired currency. Generation of this additional trade-derived data was done by taking total imports provided by the ADB tables, and estimating the component sourced from the alternate country using publicly available trade data from the United Nations (UN) Comtrade database. Commodities reported in the trade database were linked to commodities being described in the input–output tables.

103.   To estimate the induced effects in the model, worker compensation and re-spending of that money were further developed within the model. While household consumption is always reported, the portion of value-added tied to worker compensation had to be estimated. National statistics on employment and wages were used to derive the necessary data. The relevant portion of imported household consumption related to each country was similarly estimated using UN Comtrade data. The result of this effort was the generation of a multiplier model capable of portraying direct, indirect, and induced impacts, with cross-border capability.

5.2 Economic Impact Assessment Results

5.2.1 Traveler Benefit Impacts

104.   This section summarizes economic benefits typically included in a formal CBA, which was not conducted but would theoretically have involved a comparison of discounted traveler benefits against the discounted resource costs of the project. However, the ROI analysis includes (i) economic growth gains that are made possible by transportation system gains and (ii) other traveler benefits to road users and to society that are not specifically priced in normal economic exchanges.

105.   Table 5 shows traveler benefits limited to travel time and vehicle operating cost savings for existing travelers.

Table 5: Traveler Benefits for Existing Travelers by Alignment in 2030 ($ million)

Item	West					Direct	East
	New Kastek Pass		Masanchi				
	Gravel	Asphalt	Gravel	Asphalt	Tunnel		
Value of travel time savings	(5.23)	(5.76)	(3.38)	(4.42)	(6.34)	(16.04)	(11.71)
Vehicle operating cost savings	(1.56)	(1.89)	(0.83)	(1.22)	(2.23)	(7.19)	(4.69)

( ) = negative.
Source: Consultant team analysis.

106.   As outlined in Chapter 4, the various alignments will generate significant increases in new demand for tourism, both domestic and international. These additional trips represent additional economic utility and new spending. While those new trips will generate increased vehicle-kilometers of travel and vehicle-hours of travel, it would be incorrect to represent this effect in the analysis as a disbenefit, as it would in a traditional CBA framework. To a certain extent, the additional trips and resulting increases in vehicle-kilometers of travel and vehicle-hours of travel represent new demand, which in a social welfare context can be recognized (under the willingness-to-pay principle) as an increase in economic welfare. In an economic impact context (which is adopted for this report), this new demand provides a positive impact on the economy, as it represents new spending on vehicles and associated services.

107.   As seen in Table 5, existing users are projected to save around $3.4 million–$16.0 million in travel time costs and $0.8 million–$7.2 million in vehicle operating costs in 2030, when the various alignment alternatives would be open to traffic. Gravel roads provide lower traveler benefits (as they allow for lower speeds) and increased vehicle operating costs.[18] The direct alignment generates the greatest savings, as it offers the shortest route between Almaty and Issyk-Kul.

108.   Unlike travel time savings, which are not directly exchanged in markets and are thus not explicitly priced, the vehicle operating cost savings *do* affect national accounting, as they reduce demand for fuel and other vehicle repair services but compensate by increasing tourism travel and spending as a result of those savings. Those changes are in fact reflected in the EIA.

109.   Many traveler benefits would accrue to tourists originating in Kazakhstan; about 45% of existing tourists that would be diverted to the alternative road originating in Kazakhstan. Another 27% are international tourists, many of whom may arrive at Almaty airport (footnote 4).[19]

110.   Over a 20-year period of operation, and assuming a 1.6% annual increase in existing trips (reflecting the ABEC Tourism Master Plan analysis), the cumulative savings in travel time at a zero discount rate would range from about $79 million to about $374 million in constant United States dollars depending on the alignment; 20-year cumulative vehicle operating cost savings would range from $19 million to $168 million.

---

18   Vehicle operating costs are about 10% higher for gravel roads than for asphalt. M. Robbins and N. Tran. 2015. *Literature Review: The Impact of Pavement Roughness on Vehicle Operating Costs.* Auburn, Alabama, United States: National Center for Asphalt Technology.
19   Issyk-Kul International Airport in Tamchy does not show the potential to serve as a major point of access to Issyk-Kul. The rest are tourists originating in countries of the Commonwealth of Independent States other than Kazakhstan.

5.2.2 Cumulative 20-Year Impacts on National and Regional Economic Activity

111.   Not all sub-routes of the western alignment are modeled with the MRIO economic model. As cost, travel time, travel demand, and traveler benefit estimates have shown, the sub-routes across New Kastek pass or Masanchi, both gravel and asphalt, are not radically different from each other. One representative sub-route through New Kastek pass (asphalt) is used to determine economic impacts for the entire group of western alignments without tunnel.[20] This leads to a total of four modeled alignments: (i) west—without tunnel: New Kastek pass (asphalt), (ii) west—with tunnel, (iii) direct, and (iv) east.

112.   A high-level summary of the cumulative economic impacts of each of the four modeled alternatives, by policy scenario, is presented in Table 6. The table focuses on value-added, one of several economic indicators generated by the MRIO economic model. Value-added is used as the most general indicator of economic impact, as total value-added is approximately equal to GDP. The impacts in Table 6 reflect the cumulative 20-year impacts from all sources of newly generated demand-driven spending, including the following:

(i)   nonrecurring road capital construction expenditures;
(ii)   recurring annual operation and maintenance expenditures for new road infrastructure;
(iii) nonrecurring changes in travel expenditures for existing and newly induced tourist travel;
(iv)   nonrecurring new hotel construction and other supporting public and private facility expenditures; and
(v)   new tourism expenditures for hotels, restaurants, and other tourism-related daily spending.

Table 6: Summary Impacts on Value-Added (Gross Domestic Product) by Country and by Alignment and Scenario, 2025–2045

Impact Type	Kazakhstan				Kyrgyz Republic			
	West—New Kastek (Asphalt)	West—Tunnel	Direct	East	West—New Kastek (Asphalt)	West—Tunnel	Direct	East
Policy scenario II, total ($ million)	620	1,404	991	1,198	1,066	2,470	2,661	2,030
Policy scenario III, total ($ million)	1,905	3,261	3,294	3,292	3,890	7,166	8,788	7,339
Policy scenario II, average annual ($ million)	31	70	50	60	53	123	133	101
Policy scenario III, average annual ($ million)	95	163	165	165	194	358	439	367
Policy scenario II, annual share of 2017 GDP (%)	0.02	0.05	0.03	0.04	0.81	1.88	2.03	1.55
Policy scenario III, annual share of 2017 GDP	0.07	0.11	0.11	0.11	2.96	5.46	6.69	5.59

GDP = gross domestic product.
Source: Consultant team analysis.

---

20   This alignment is shown to yield the highest travel demand and traveler benefits (Table 5).

113.   Three important observations stand out:

(i)   Over a 20-year period, the cumulative GDP increase in Kazakhstan would range from $0.6 billion to $3.3 billion, depending on the alignment and policy scenario; the equivalent gains for the Kyrgyz Republic would range from $1.1 billion to $8.8 billion.

(ii)   On an average annual basis, the relative economic gain in Kazakhstan would range from 0.02% to 0.11% of total annual GDP (based on 2017 GDP levels); equivalent gains in the Kyrgyz Republic would range more dramatically from 0.8% to 6.7%.

(iii) The three alignments that include a tunnel produce economic impacts of a similar magnitude over the 20-year observation period. However, the individual components (shown for three alignments in Table 7) offer additional insights. Within the 20-year period, construction-related impacts, covering 7–9 years for these tunnel alignments, represent a significant share of the total impacts. The direct alignment produces the greatest economic impact from additional tourism demand because of the higher travel volumes. Because the construction costs represent nonrecurring impacts, the direct alignment would clearly overtake the west—tunnel and the eastern alignment over a longer time frame.

114.   Further details of cumulative results, including a breakdown of value-added contribution by source of expenditure and direct, indirect, and induced impacts, are shown in Table 7. From this perspective, the added construction, tourism demand, and induced new trips all contribute to more value-added in the economy, while the savings in fuel and other travel costs for existing travelers actually represent a loss of spending. This is accounted for in Chapter 6, where it is recognized as a social welfare benefit.

Table 7: Cumulative Value-Added Impacts by Source of Addition to Final Demand for Three Representative Road Alternatives and Scenarios

Cumulative Value-Added (GDP) Impacts (2025–2045), New Kastek (Asphalt), Scenario II

Impact Type	Kazakhstan				Kyrgyz Republic				Total
	Direct	Indirect	Induced	Total	Direct	Indirect	Induced	Total	Total
Highway construction ($ million)	53	47	45	**145**	1	1	1	**4**	**149**
Travel cost savings: existing ($ million)	(21)	(8)	(10)	**(39)**	(1)	0	0	**(1)**	**(40)**
Travel cost savings: induced ($ million)	38	15	17	**70**	2	0	1	**2**	**72**
Tourism demand ($ million)	151	124	169	**444**	451	240	370	**1,061**	**1,505**
Hotel construction ($ million)	0	0	0	**0**	0	0	0	**0**	**0**
Supporting construction ($ million)	0	0	0	**0**	0	0	0	**0**	**0**
Total impact ($ million)	220	178	221	620	453	241	372	1,066	1,686
Average annual impact ($ million)	11	9	11	31	23	12	19	53	84
Share of 2017 GDP (%)				0.02				0.81	

continued on next page

Table 7 *continued*

Cumulative Value-Added (GDP) Impacts (2025–2045), West Kastek (Asphalt), Scenario III

Impact Type	Kazakhstan				Kyrgyz Republic				Total
	Direct	Indirect	Induced	Total	Direct	Indirect	Induced	Total	
Highway construction ($ million)	53	47	45	**145**	1	1	1	**4**	149
Travel cost savings: existing ($ million)	(21)	(8)	(10)	**(39)**	(1)	0	0	**(1)**	(40)
Travel cost savings: induced ($ million)	129	51	60	**240**	5	1	3	**8**	249
Tourism demand ($ million)	518	427	580	**1,525**	1,548	825	1,271	**3,643**	5,167
Hotel construction ($ million)	0	12	17	**29**	73	54	69	**196**	225
Supporting construction ($ million)	0	2	3	**6**	15	11	14	**39**	45
Total impact ($ million)	**679**	**532**	**695**	**1,905**	**1,641**	**891**	**1,357**	**3,890**	**5,795**
Average annual impact ($ million)	34	27	35	95	82	45	68	194	290
Share of 2017 GDP (%)				0.07				2.96	

Cumulative Value-Added (GDP) Impacts (2025–2045), Direct, Scenario III

Impact Type	Kazakhstan				Kyrgyz Republic				Total
	Direct	Indirect	Induced	Total	Direct	Indirect	Induced	Total	
Highway construction ($ million)	17	49	63	**129**	213	158	200	**571**	700
Travel cost savings: existing ($ million)	(3)	(4)	(7)	**(14)**	(56)	(7)	(25)	**(88)**	(103)
Travel cost savings: induced ($ million)	4	5	9	**18**	70	9	32	**111**	129
Tourism demand ($ million)	998	823	1,117	**2,938**	2,983	1,589	2,449	**7,021**	9,959
Hotel construction ($ million)	15	73	97	**186**	364	271	343	**978**	1,164
Supporting construction ($ million)	3	15	19	**37**	73	54	69	**196**	233
Total impact ($ million)	**1,034**	**961**	**1,299**	**3,294**	**3,647**	**2,074**	**3,067**	**8,788**	**12,083**
Average annual impact ($ million)	52	48	65	165	182	104	153	439	604
Share of 2017 GDP (%)				0.11				6.69	

( ) = negative, GDP = gross domestic product.
Source: Consultant team analysis.

115.   While direct impacts account for only about 30% of total impacts in Kazakhstan, this share is about 40% in the Kyrgyz Republic. The reason for this is a stronger spillover effect of additional economic activity in Issyk-Kul, leading to a broader impact in Kazakhstan through purchases of goods and services that will go to serve tourism needs in the Kyrgyz Republic. Examples of such goods and services, derived from the statistics of trade between the two countries, may be in food and agricultural goods, as well as gasoline.[21] Demand for

---

21   United Nations. UN Comtrade Database (accessed 31 October 2020).

agricultural goods in the Kyrgyz Republic comes from added activity of households and businesses and includes importing raw agricultural goods (e.g., cereal grains and animal by-products) as well as more finished products (e.g., tobacco, beverages, and flour). Similarly, much of Kazakhstan's export of gasoline is imported by the Kyrgyz Republic. Because of the interrelatedness of trade and supply chains, a portion of demand for goods and services is met through trade with neighbors, which influences the employment impacts shown in Chapter 5, section 5.2.3.

116. Impacts on wage income and employment follow similar general patterns as the value-added impacts (Table 7). In the exemplary combinations of alignments and scenarios chosen for Table 8, the annual numbers are less than in Table 7. This is because in Table 8 the single year shown, 2045, is post-construction, and thus construction impacts are not included in the annual number. Instead, they are in Table 7, which shows average annual figures over the combined construction and post-construction operations period.

Table 8: All Economic Indicators for Two Representative Combinations of Alignments and Scenarios for 2045

Western Alignment, New Kastek (Asphalt), Policy Scenario II, 2045

	Kazakhstan				Kyrgyz Republic				
Item	Employment (No. of jobs)	Income	Value-Added ($ million)	Output	Item	Employment (No. of jobs)	Income	Value-Added ($ million)	Output
Direct	972	4.5	11.6	20.1	Direct	12,269	28.2	29.7	71.7
Indirect	474	2.9	9.2	15.8	Indirect	2,601	12.2	15.8	40.3
Induced	699	3.6	12.1	19.9	Induced	4,742	19.5	24.4	61.1
Total	2,146	11.0	32.9	55.8	Total	19,612	59.9	69.8	173.1
Share of 2017 GDP (%)	0.02		0.02	0.02	Share of 2017 GDP (%)	0.82		1.06	1.08
2017 GDP	8,585,153		145,489	244,844	2017 GDP	2,377,700		6,567	16,084

Direct Alignment, Policy Scenario III, 2045

	Kazakhstan				Kyrgyz Republic				
Item	Employment (No. of jobs)	Income	Value-Added ($ million)	Output	Item	Employment (No. of jobs)	Income	Value-Added ($ million)	Output
Direct	7,974	35.2	91.2	157.6	Direct	113,186	264.8	279.1	681.4
Indirect	4,056	24.1	76.9	132.9	Indirect	24,700	116.4	150.3	383.0
Induced	5,974	31.4	104.0	171.2	Induced	44,766	184.4	229.9	577.1
Total	18,005	90.7	272.1	461.6	Total	182,651	565.6	659.3	1,641.5
Share of 2017 GDP (%)	0.21		0.19	0.19	Share of 2017 GDP (%)	7.68		10.04	10.21
2017 GDP	8,585,153		145,489	244,844	2017 GDP	2,377,700		6,567	16,084

GDP = gross domestic product.

Note: Income, value-added, and output cannot be summed. Income is included in value-added, and value-added is included in output.

Source: Consultant team analysis.

117.   Importantly, the number of jobs created in the Kyrgyz Republic is higher than they would be in many other more developed economies, including Kazakhstan, because wages are relatively low in the Kyrgyz Republic. However, the model includes adjustments to the employment numbers made afterward based on the assumption that by 2045 the Kyrgyz Republic's economy will be considerably more developed, and average wages and worker productivity will have increased. With those increases, the total number of jobs created (i.e., the ratio of jobs to output and value-added) will fall relative to what might be expected given current economic data, and the rate of job creation will be less in the Kyrgyz Republic than at present. Improvements in worker productivity and skill, consistency of employment attendance, and a reduction in employment in informal sectors will all make possible such a reduction in the gross number of employees while also raising household incomes of new job holders. These adjustments also make the ratio between additional visitors and newly created tourism jobs more comparable to reported numbers from other countries in the region.[22]

118.   The effects of traffic flow shifts are not considered in the modeled economic impacts. Businesses along the existing roads through Korday and Karkyra will lose revenue when the alternative road attracts shares of their traffic volumes. The consultant team conducted a field survey to assess the number of businesses that could be affected. Businesses were considered in two categories: (i) restaurants, cafes, or bars were counted on the basis of provided parking spots for customers; and (ii) gas stations were counted by the number of gas pumps.[23] It was concluded that the businesses along the existing routes would lose 6%–8% of revenue in the case of one of the western alignments and 11%–12% with the direct and eastern alignments. However, most or all of the losses along existing routes will be offset by revenue for new businesses along the alternative road, meaning there may ultimately be little net impact of these shifts, though they are likely to represent gains and losses for different people.

5.2.3 Detailed Employment Impacts

119.   Viewing the impacts of the highway project on employment and wage income by industry in a specific future year, in this case 2045, provides additional insight. That year was chosen, as it allows for more than 10 years of operation for each alignment and thus captures post-construction period impacts consistently across all alignment alternatives. While measuring the economic impact focus in large part on value-added (a proxy for GDP increase) as the most comprehensive economic indicator, policy makers and stakeholders are particularly concerned with job and labor income creation, as these best indicate levels of economic well-being and satisfaction for the broadest layer of beneficiaries in an economy. Value-added includes not only labor income, but proprietor income and profit margins.

120.   Table 9 summarizes employment and labor income impacts for the western alignment through New Kastek pass (asphalt) under policy scenario III. For other scenarios, the basic distribution of impacts by economic sector will follow the same general pattern as seen in the table.

121.   As seen in Table 9, both countries will gain significant numbers of jobs and associated worker income. Because most of the newly created direct jobs will be in the hospitality industry (e.g., hotels and restaurants), and because most of those jobs will be created in the Kyrgyz Republic proximate to the lake, considerably more

---

22   While 50 tourism jobs per 1,000 arrivals were reported for Armenia, 69 for Azerbaijan, and 22 for Georgia, the adjusted job numbers for Issyk-Kul imply a ratio of about 40 created jobs per 1,000 additional arrivals. World Travel & Tourism Council. 2019. *Travel & Tourism, Benchmarking Research Trends 2019.* London. World Travel & Tourism Council. 2013. *Travel & Tourism, Economic Impact 2013 Turkey.* London. Forbes Georgia. 2018. *Contribution of tourism to the economy of Georgia and other countries of the world.*

23   Reduction of traffic volumes was estimated by segment and weighted by the number of restaurants, cafes, or bars and gas stations along the segment. In total, about 3,000 parking spots for customers and 1,000 gas pumps were counted along the approximately 930 kilometers of existing road on both routes.

Table 9: Job and Income Impact by Industry for West—New Kastek Alignment, Scenario III, 2045

Industry	Employment (No. of jobs)		Income ($ million)	
	Kazakhstan	Kyrgyz Republic	Kazakhstan	Kyrgyz Republic
Services	941	3,159	6.0	19.2
Transport	1,489	13,540	8.6	38.9
Hospitality	2,143	33,044	8.1	68.6
Trade	840	7,136	6.2	34.7
Construction and utilities	232	1,483	1.6	8.1
Manufacturing	290	875	2.4	7.3
Agriculture and food	1,539	9,565	5.6	35.5
Total	**7,474**	**68,802**	**38.5**	**212.3**

Share of Total Impact, by Industry (%)

Industry	Employment		Income	
	Kazakhstan	Kyrgyz Republic	Kazakhstan	Kyrgyz Republic
Services	13	5	16	9
Transport	20	20	22	18
Hospitality	29	48	21	32
Trade	11	10	16	16
Construction and utilities	3	2	4	4
Manufacturing	4	1	6	3
Agriculture and food	21	14	14	17
Total	**100**	**100**	**100**	**100**

Source: Consultant team analysis.

jobs will be created in that country. It may present challenges to fill these jobs with trained workers, a constraint not considered in the calculations.

122.   In addition to the notable gains in hospitality businesses, significant increases are modeled in agriculture and food production (to feed the added visitor base) and air and ground transport (to serve them). There are also notable impacts on supporting more manufacturing, wholesale, and retail trade. This will be true in both countries.

123.   Additional information on the distribution of impacts by country is found in Figure 13. In particular, a larger share of total job growth in the Kyrgyz Republic will be in hospitality, while in Kazakhstan it will be in agriculture/food, transport, and services. However, both countries will see significant boosts to their agriculture and food sectors, and cross-border supply chains will benefit both countries.

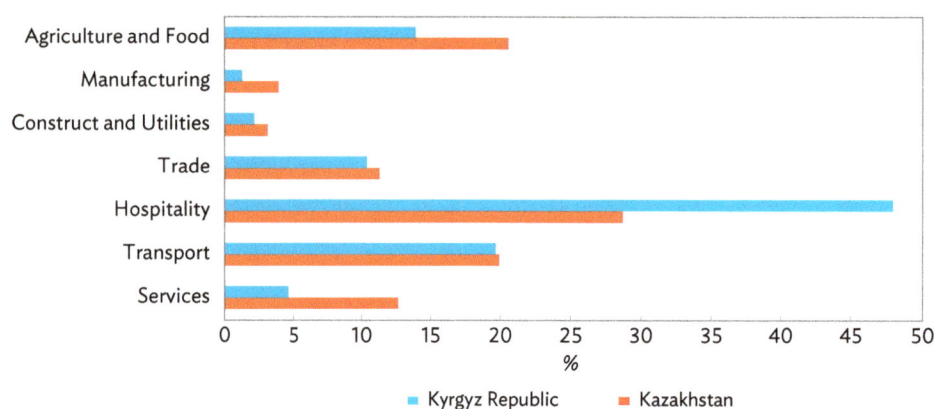

Figure 13: Distribution of Created Jobs across Industries for the Western Alignment through New Kastek Pass (Asphalt) (Representative for All Alignments)

Source: Consultant team analysis.

5.2.4 Fiscal Impacts

124. The increases in economic outcomes described in paras. 104–123 (e.g., increased output, growth in final and intermediate business sales, personal and business wages and other forms of compensation income, new commercial development) generate tax revenue increases for both countries, and they may also generate local and oblast-level tax impacts. The consultant team has estimated the national tax revenue impacts based on data relating to overall tax collection within each country, by individual tax category.

125. These estimates are based on highly aggregated information, rather than on a ground-up assessment of individual tax bases and tax rates. More specifically, tax impacts are estimated based on the ratio of annual tax revenues collected (by country) to the corresponding year's annual economic output for that country. Applying these ratios to the estimated gains in overall output estimated by the MRIO economic model produces the fiscal impacts shown in Table 10 by country. Again, these findings are reported for the representative year of 2045 for the western alignment through New Kastek pass (asphalt) under policy scenario III conditions.

126. In interpreting these findings, it is important to note that the tax revenue effects are reported from the standpoint of the government entity *collecting* the tax; it does not reflect the final incidence of the tax, or who pays the tax. For example, while consumption taxes shown in Table 10 indicate significant gains in consumption-based taxes in the Kyrgyz Republic, a significant (but not estimated) share of those taxes will be paid by tourists who are not Kyrgyz residents. Cross-border tax adjustments between the two countries that may be in effect are not considered.

127. As seen in Table 10, in 2045 tax revenues will increase in Kazakhstan by about $9 million per year (in 2019 United States dollars) for this alignment and policy combination. Consumption-based taxes collected by Kazakhstan will increase by $5 million, while consumption-based taxes in the Kyrgyz Republic will grow by $12 million for this alignment and scenario combination. The amounts collected (i.e., the tax revenue gains) are expected to be higher in the Kyrgyz Republic since a large share of new consumption by tourists and workers will occur in the Kyrgyz Republic (at stores, restaurants, and other businesses). The percentage gain in overall tax proceeds in the Kyrgyz Republic will be even larger since the Kyrgyz tax base is smaller. More detailed information can be found in the Economic Impact Assessment (Supplementary Document that can be downloaded on http://www.almaty-bishkek.org).

Table 10: Tax Impacts by Country for the West—New Kastek Alignment, Scenario III, 2045

		Share (%)	Amount ($ million)
Kazakhstan	**Output from EIA in 2045** ($ million)	200.8	
	Tax Type		
	Personal income tax for residents	0.533	1.04
	Personal income tax for nonresidents	0.004	0.01
	Profits tax	1.101	2.16
	Property tax	0.214	0.42
	Land tax	0.012	0.02
	Consumption taxes	2.557	5.00
	Other	0.699	0.00
	Total ($ million)		8.7
	Percentage of national tax revenue (%)		0.04
Kyrgyz Republic	**Output from EIA in 2045** ($ million)	653.3	
	Tax Type		
	Personal income tax for residents	0.444	2.73
	Personal income tax for nonresidents	0.094	0.58
	Profits tax	0.208	1.28
	Property tax	0.076	0.47
	Land tax	0.061	0.37
	Consumption taxes	1.982	12.20
	Other	0.688	0.00
	Total ($ million)		17.6
	Percentage of national tax revenue (%)		0.06

Sources: Tax information from Kazakh State Revenue Committee and the Kyrgyz State Tax Service; consultant team analysis.

6

Return on Investment

6.1 Definition of Metrics for Return on Investment

128.   For ADB as well as for the governments of Kazakhstan and the Kyrgyz Republic, it is desirable to not only achieve significant economic growth outcomes (Chapter 5), but also to ensure efficiency in terms of a reasonable ROI. This chapter focuses on the measurement of ROI from the perspective of ADB and its goal of helping Kazakhstan and the Kyrgyz Republic to achieve greater prosperity and economic development.

129.   Traditionally, the efficiency of transportation project funding has been examined via cost–benefit analysis (CBA). This methodology applies the concept of a discounted present value measurement to compare a stream of project benefits occurring over time with a stream of investment costs occurring with different timing. CBA is most frequently used to judge the full social welfare gains from investments, particularly where there are likely to be major benefits in areas where markets either do not exist or are only secondary (e.g., air emissions are social benefits that are not priced, or only partially priced, in carbon trading markets).

130.   While the present value concept is quite appropriate, the practice of CBA for transportation projects is usually focused on transportation system efficiency, which sums the social value of time, cost, and safety benefits for travelers. With this perspective, additional economic impacts on attracting inward investment, economic growth, and prosperity to a region are rejected as being "spatial shifts" of benefits that would otherwise occur elsewhere.

131.   But in this case, the interest of ADB is on attracting inward investment and economic growth, prosperity, and a higher quality of life to this region. Furthermore, the transportation system efficiency view of time and cost savings is in this case less relevant, as there are a limited number of travelers currently traveling between Almaty and Issyk-Kul via the existing route. As a consequence, there are limited time and cost savings for current travelers, and much more interest in the cost-effectiveness of outcomes that represent the attraction of additional visitor spending, investment, construction, and recreation to the region.

132.   While significant public resources would be committed to the infrastructure development itself (at least in the short run, either through grants, direct loans, or loan guarantees), the economic benefits would primarily be obtained by private market participants, including workers and, to a much larger extent, businesses and business owners in the tourism and real estate development industries. In this case, private investors most frequently turn to internal rate of return analysis as the best and most telling indicator of economic and financial viability of a project. Because of the mixed nature of public investment and private returns, the blended economic internal rate of return (EIRR) approach is used here as the best measure of economic viability. This methodology has been adapted to reflect certain key private sector assumptions, such as the use of a relatively short (10-year) road operation time frame of long-term annual economic yield to restrict the analysis parameters and make it more comparable to private sector investment decision-making.

133.   Therefore, instead of using the traditional view of CBA, the consultant team adopts the spatial perspective of the ABEC region as a part of the Kazakh and the Kyrgyz economies as its area of interest, and the team applies a hybrid form of ROI analysis that counts benefits as including (i) GDP growth for the two countries that are enabled by the project (Chapter 5, section 5.2.2) and (ii) traveler welfare benefits (such as time and cost savings) (Chapter 5, section 5.2.1) that are not already captured in measures of GDP growth. In this case, the value of travel time savings for existing travelers (which was not counted as a driver of the GDP growth calculation) and the value of travel cost savings (which was counted as spending reduction in Chapter 5, section 5.2.2) are counted.

134.   This perspective represents a comprehensive form of ROI analysis that brings together both elements of social welfare analysis and economic impact analysis to gain a richer understanding of what public investments can deliver to society in general as well as to national income. It also provides for a more nuanced inclusion of benefits that cannot be readily quantified but are indeed tangible, such as national economic outlook, impacts on business climate, greater economic choice, and more balanced and equitable economic development.

135.   This chapter provides such an expanded outlook. Three primary indicators are presented in this section:

   (i)   **An expanded economic internal rate of return (section 6.2).** This presents a comparison of the combined economic impact and traveler benefits of alternatives relative to their investment costs. This is referred to as an EIRR even though different concepts may be known under this term. An internal rate of return represents the discount rate that equates project benefits with project costs, and reflects the actual rate of return, which can be compared with the opportunity cost of capital in a given country, to determine if the project exceeds the standard rate of return required to generate economic investment. In this case, care is exercised to eliminate double counting; for example, while construction spending generates an economic stimulus and an increase in final demand in the construction sector, it also entails an opportunity cost. As a result, in this ROI framework, the direct costs of construction, as well as their indirect and induced effects, are not included on the benefits side of the ROI calculation. Instead, the direct expenditures are registered as costs.

   (ii)   **Financial internal rate of return (section 6.3).** This presents a comparison of revenue from various sources of alternatives relative to cost of capital and returning costs. This is not exactly an economic indicator; instead it expresses financial viability of an investment. However, this measure is still considered important to showcase one specific way of (viable) financing of the project.

   (iii) **Multi-criteria rating (section 6.4).** This presents a summary template that rates each alternative across a range of quantitative and qualitative indicators. Ratings range from 1 to 5, with 5 being the most beneficial. This technique relies on qualitative ratings to supplement the EIRR by providing a way to consider additional benefits and goals that cannot be directly monetized, such as environmental benefits and quality of life (experiential value). With this approach, the alternative alignments can also be compared in terms of traveler benefits, GDP, employment, new business investment, and EIRR. In addition, the financial feasibility and job creation effect of those alternative scenarios can be separately rated.

6.2 Economic Internal Rate of Return and Economic Break-Even Point

136.   The EIRR as defined in section 6.1 is shown in Table 11 for each alignment, by scenario. The EIRRs range from 12.2% to 69.0% depending on alignment and policy scenario, which means they are well above ADB's minimum required EIRR of 9%.[24]

Table 11: Expanded Economic Internal Rate of Return, by Alignment and Policy Scenario
(%)

Alignment	Scenario II	Scenario III
West—New Kastek (Asphalt)	36.5	69.0
West—Tunnel	13.7	24.6
Direct	12.2	23.5
East	13.7	28.3

Note: To avoid double accounting, gross domestic product gains exclude gains from new road construction.
Source: Consultant team analysis.

137.   The reason for the EIRR being so high may be that it does not only reflect a benefit provided by the alternative road but rather the benefit from removing a major access obstacle by building the road. It is also a regional benefit, as it explicitly counts benefits from induced tourism spending attracted into this region from other countries or continents. Given the situation—a large economic center at one end, and an attractive tourism area at the other—the new road would largely serve a demand that can be assumed to be latently present but that cannot be satisfied within the ABEC region as there are no similarly attractive destinations close enough to entice travelers to undertake trips for shorter stays like weekends. Similarly, for residents of the Issyk-Kul area, there may already be a desire to travel more frequently to Almaty, but travel times are prohibitively long.

138.   Tourism forecasts for scenario III are high and represent the result of far-reaching measures like improvements to other tourism infrastructure or changes in policies. As shown in Chapter 4 (Figures 7–10), even though the road would benefit from a scenario III situation and see high travel demand, it is not assumed that the road by itself would reach demand levels as described in scenario III of the ABEC Tourism Master Plan.

139.   Under policy scenario II, the exemplary western alignment without tunnel—New Kastek pass (asphalt)—exhibits a high EIRR, as its annual economic development impacts are high relative to the other alternatives when few supportive policies are in place (i.e., scenario II). However, under both policy scenarios, all alignment alternatives score well on EIRR, with variations depending on costs relative to benefits.

140.   The economic break-even point is defined as the point where net present value (NPV) of economic benefits minus costs equals zero at a 9% discount rate (which corresponds to the threshold used by ADB for economic viability). Figure 14 shows the level of travel demand that is required to reach the economic break-even point. For all four considered alignments,[25] conditions with a moderate growth of tourism flows—as described in policy scenario II in the ABEC Tourism Master Plan (Chapter 4, section 4.1)—are sufficient to reach economic viability.

---

24   ADB. 2017. *Guidelines for the Economic Analysis of Projects.* Manila.
25   See Paragraph 111.

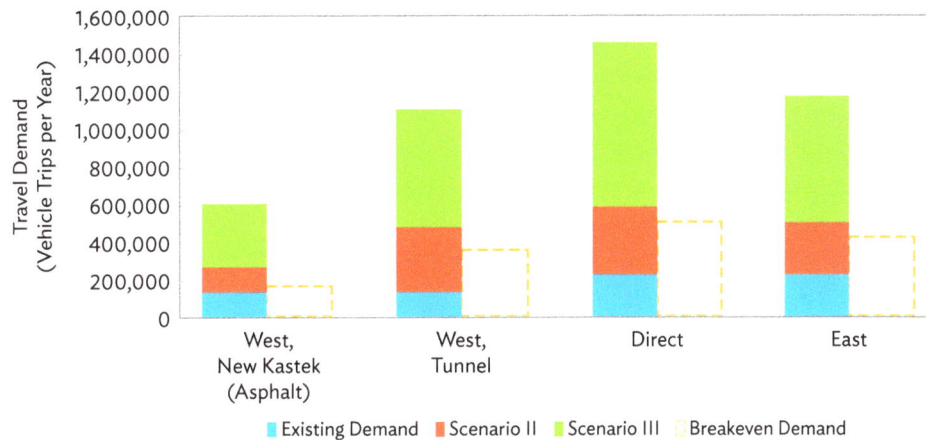

Figure 14: Economic Break-Even Points—Travel Demand Required to Reach Economic Viability

6.3 Financial Internal Rate of Return and Financial Break-Even Point

141.   While this study focuses most of its attention on EIA and EIRR calculations, it also presents an initial rough estimate of financial internal rate of return. This demonstrates an approximated financial viability for a scenario, in which the alternative road is implemented as a toll road under a public–private partnership (PPP) scheme. Toll rates are assumed to correspond to half the out-of-pocket cost savings experienced by travelers when diverted from existing routes to the alternative road. This corresponds to one of the two settings for toll rates quantified in Chapter 7.4.

142.   The following assumptions are used for this approximation:

(i)   Toll rates are 50% of out-of-pocket cost savings, resulting in toll rates ranging from $3.70 to $18.73 per trip depending on the alignment.
(ii)   Road construction cost is a sunk cost. It is not feasible to amortize capital cost through tolling at rates deemed reasonable. Neither a capital facilities charge nor any tourism tax to amortize the debt over time is considered in this approximation.
(iii)   Capital cost of tolling infrastructure is $5 million for all alignments.
(iv)   Tolling operation cost is $0.5 million per year for all alignments.
(v)   Toll revenue equals the travel demand by scenario multiplied by the toll rate by alignment.
(vi)   Roadside service revenue equals $0.5 million per year for all alignments.
(vii)   Years of operation considered after the construction phase will be 20 years.

143.   The results in Table 12 show how financial outcomes depend on the alignment and the travel demand (policy scenarios II and III).

Table 12: Financial Internal Rate of Return, by Alignment and Policy Scenario

Alignment	Scenario II FIRR	Scenario III FIRR
West—New Kastek (Gravel)	6.5%	42.5%
West—New Kastek (Asphalt)	12.5%	59.5%
West—Masanchi (Gravel)	NPV < 0	NPV < 0
West—Masanchi (Asphalt)	NPV < 0	13.0%
West—Tunnel	NPV < 0	59.0%
Direct	33.0%	>100%
East	NPV < 0	>100%

FIRR = financial internal rate of return, NPV = net present value.
Source: Consultant team analysis.

144.   In this financing scenario under the assumptions shown in para. 142, only three of the seven alignments appear to be financially viable,[26] with a moderate increase of travel demand corresponding to policy scenario II from the ABEC Tourism Master Plan. Three sub-routes of the western alignment and the eastern alignment require a higher travel demand (policy scenario III) to reach viability, while one alignment, West—Masanchi (asphalt) does not generate enough toll revenue to cover the cost, even under policy scenario III with high travel demand.

145.   The financial break-even point is defined as the point where NPV of the cash flow over the course of 20 years equals zero at a discount rate that corresponds to the weighted average cost of capital, which is assumed to be 4% in constant United States dollars. Figure 15 shows the level of travel demand necessary for each alignment to reach a positive NPV at a 4% interest rate.

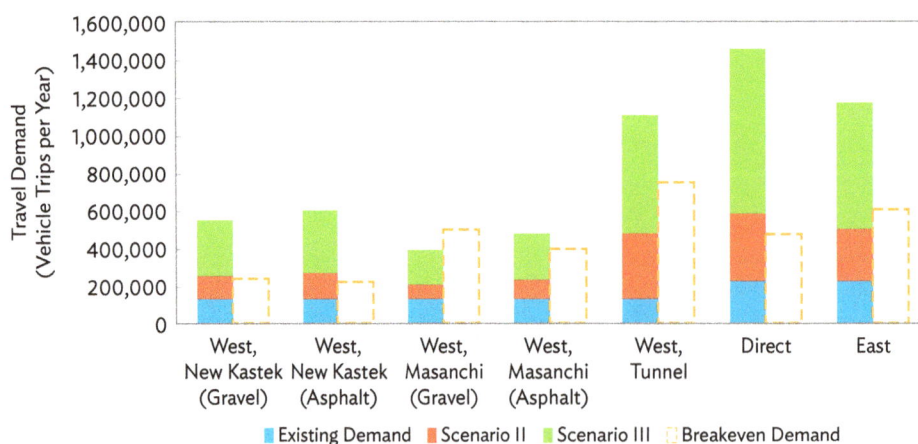

Figure 15: Necessary Travel Demand for Financial Break-Even

Source: Consultant team analysis.

---

26   In order to be financially viable, an alignment has to yield an NPV greater than zero.

146.   The West—Masanchi (gravel) alignment is not financially viable under any policy scenario. It would require much higher travel demand to generate sufficient toll revenue. Three alignments require a strong increase in travel demand that surpasses the forecast for policy scenario II from the ABEC Tourism Master Plan. The direct alignment yields the highest toll revenue as it would justify the highest toll rate and is also expected to attract the strongest demand. The western alignment through New Kastek pass (gravel or asphalt) also appears to be financially viable even with a more moderate demand increase.

6.4 Multi-Criteria Rating

147.   Each criterion is scored on a scale of 1 to 5, where 5 means very favorable and 1 the least favorable. Scores cannot be summed, as weightings have not been determined for each criterion.[27] The scores are meant to signal the relative effects of an alignment. Again, the sub-route across New Kastek pass (asphalt) represents the western alignments without tunnel, which, relative to each other, are assessed similarly (Table 13).

Table 13: Multi-Criteria Rating (Scenarios II and III Collectively)

Alignment	Traveler Benefits	GDP	Employment and Income	Economic Viability (EIRR)	New Business Investment	Environment	Experience Value	Financial Viability (FIRR)
West—New Kastek pass (asphalt)	2	3	3	5	2	4	4	4
West—Tunnel	3	4	4	4	3	4	3	2
Direct	3	5	5	4	5	3	5	5
East	3	4	4	4	4	3	5	3

EIRR = economic internal rate of return, FIRR = financial internal rate of return, GDP = gross domestic product.
Source: Consultant team analysis.

148.   To complete the multi-criteria rating, we introduce the environment and experience value aspects of the alternative road between Almaty and Issyk-Kul, in addition to the economic and financial metrics. They involve the following:

(i)   **Environment.** Two main factors are considered within this criterion.
   (a)   **Air pollution and carbon dioxide emissions.** Existing traffic is diverted to the alternative road, which reduces the vehicle-miles traveled and emissions. On the other hand, new induced traffic will lead to increased emissions. As induced demand will be a major part of traffic on the alternative road, the negative impact weighs heavier.
   (b)   **Landscape and nature.** All alignments cross mountain areas, which are still mostly untouched. The level of interference is relatively high. The impact is stronger for the direct and eastern alignments, crossing two mountain ridges (albeit one of them in a tunnel) and touching Chon-Kemin (direct alignment) and Kolsay Lakes (eastern alignment) national parks or other natural resources.
(ii)   **Experience value.** As they cross remote mountain areas, all alignments provide scenic views and a mountain experience very different from what most travelers know. The experience is even stronger with the direct and eastern alignments, which climb to high altitudes and traverse national parks. Tunnels may be seen as a negative feature of some alignments, as many travelers may not be used to crossing tunnels. However, it is assumed throughout this study that tunnels should be illuminated, ventilated, and safe.

---

27   Summing the scores would in fact be tantamount to an equal weight for all criteria, something that cannot simply be assumed.

7

Financing and Funding

7.1 Introduction

149.   Different ways to finance an alternative road between Almaty and Issyk-Kul are laid out in this chapter. Besides financing of the construction costs with loans and grants by an international development finance institution like ADB, dedicated taxes or a PPP model may also be considered to cover operation and maintenance costs.

150.   As two countries are involved, apart from ways to finance the road, it should also be determined how to split the cost.

7.2 Traditional Financing of Investment Cost

151.   Kazakhstan is part of ADB's Group C of developing member countries and can therefore borrow from ordinary capital resources at near-market terms only, whereas the Kyrgyz Republic, as member of Group A, is eligible for concessional assistance. Various ways of splitting the construction cost are conceivable, although they deliver different results.

152.   **Cost split by territory.** If cost was split on the basis of road construction costs being incurred in each national territory, the countries' shares would be radically different depending on the alignment. While the Kyrgyz portion would be very small for the western alignments without tunnel, the cost would be more evenly split for the western alignment with tunnel and for the eastern alignment. Still, even though almost all the cost would be on the Kazakh side for the western alignments, the absolute cost for Kazakhstan would be smaller than with the more expensive western alignment with tunnel and the eastern alignment (Figure 16). The direct alignment, though, entailing the highest construction cost, is located almost entirely on the Kazakh side, leading to an unequal split of the cost.

153.   **Cost split by social benefits.** Users currently traveling between Almaty and Issyk-Kul will save travel time and cost with an alternative road. Residents of Kazakhstan and the Kyrgyz Republic will benefit to different extents according to their share among the travelers.[28] This way to split the construction cost leads to nearly the same result for all alignments (Figure 17).

154.   **Cost split by economic impacts.** As the alternative road may be perceived as a tool to induce economic development at both ends of the road, it may also be seen as a fair way to split the cost along the lines of economic impacts expected on both sides of the border. Of the value-added accruing for the period 2025–2045, about 70% is projected to occur in the Kyrgyz Republic and about 30% in Kazakhstan. These shares

---

28   According the ABEC Tourism Master Plan, 45% of travelers originate from Kazakhstan, 27% from the Kyrgyz Republic, and 28% are international tourists.

Figure 16: Construction Cost Split by Territory
($ million)

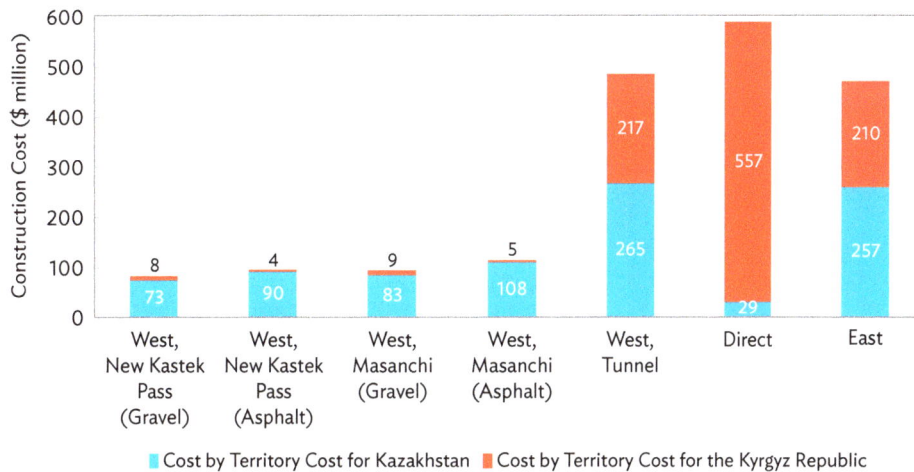

Cost by Territory Cost for Kazakhstan   Cost by Territory Cost for the Kyrgyz Republic

Source: Consultant team analysis.

Figure 17: Construction Cost Split by Social Benefits
($ million)

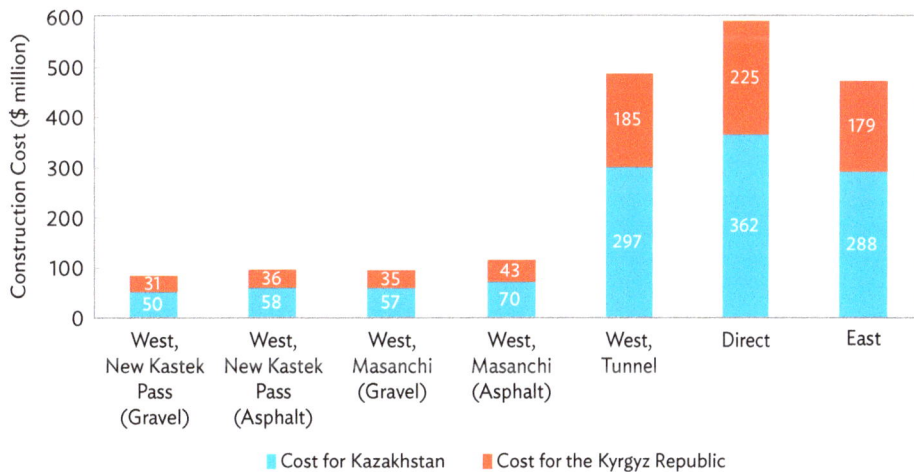

Cost for Kazakhstan   Cost for the Kyrgyz Republic

Source: Consultant team analysis.

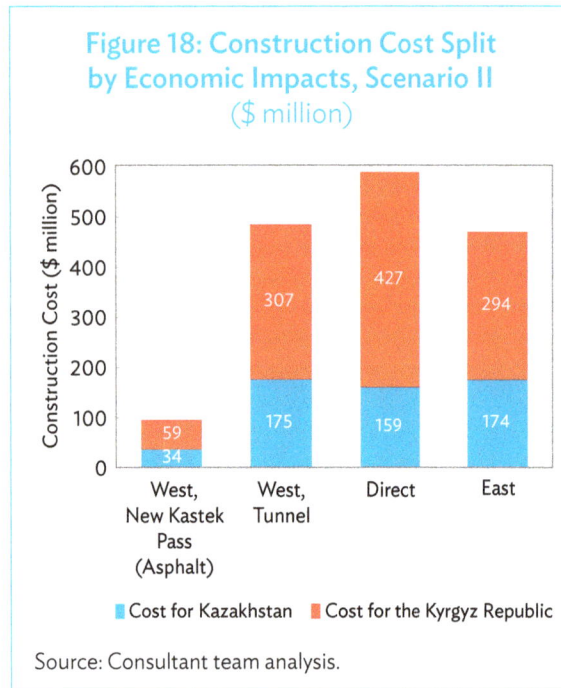

Figure 18: Construction Cost Split by Economic Impacts, Scenario II ($ million)

Source: Consultant team analysis.

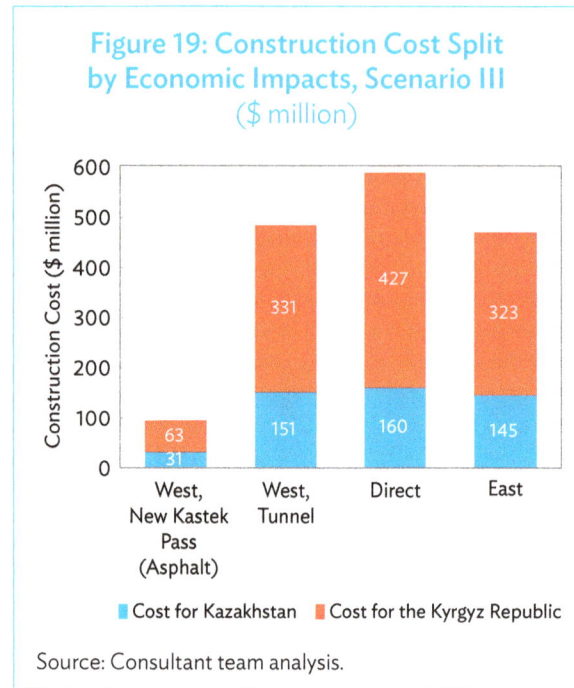

Figure 19: Construction Cost Split by Economic Impacts, Scenario III ($ million)

Source: Consultant team analysis.

slightly vary between alignments and scenarios. Figure 18 and Figure 19 show the splits for the four alignments included in the EIA.

7.3 Funding of Operation and Maintenance Costs through Dedicated Taxes

155.   Instead of using general government funds, dedicated taxes could be raised to provide the means for funding operation and maintenance of the alternative road. Those taxes would target the beneficiaries of the economic impacts enabled by the alternative road. As the road's principal purpose is to serve travelers to increase tourism demand and thereby drive economic development, a tourist tax could be imposed.

156.   It is estimated that an accommodation tax of $1.75 per person per night would be sufficient to cover operation and maintenance costs for the more expensive alignments that include tunnels.[29] This would presuppose, though, that tourism demand would be growing fast along the path described in the ABEC Tourism Master Plan's policy scenario III.

7.4 Public–Private Partnership Model

157.   In a Public–Private Partnership (PPP) model, government entities and private partners allocate the tasks, obligations, and risks in a way that recognizes the relative advantages each partner has. The government's contribution to a PPP may take the form of capital for investment or in-kind contributions that support the PPP. Private entities may also contribute investment capital, but they mainly bring in their expertise in management

---

29   This corresponds to the accommodation tax introduced by Turkey in 2020 (TL12 per person per night for four-star hotel accommodations).

and operations. For-profit-oriented thinking may provide more efficient and user-friendly operations of the alternative road.

158.   PPP model development would in this case involve two governments. Examples of successful binational PPP models exist in other countries.[30] Additionally, the collaboration between the governments of Kazakhstan and the Kyrgyz Republic in implementing new BCPs may serve as examples.

159.   It is approximated what the toll rate would have to be if a private company were to cover its expenses for operation and maintenance (without any profit) (Table 14).[31] While the average toll could be in a range below $10 per vehicle for the western alignments without tunnel, it would have to be higher for alignments with tunnel, if the policies considered in scenario III are not adopted and the number of visitors grows only moderately.

160.   A different approach to estimating toll rates is to assume that they correspond to half of the travelers' out-of-pocket costs for vehicle operation (Table 15). This would mean that travelers pass on half of their monetary benefits to the road operator, while they still benefit from considerably lower out-of-pocket costs than on the existing routes.

161.   A comparison between the two approaches (Table 14 and Table 15) show that toll rates are in a similar general range. Toll rates for alignments without tunnel can be expected to be under $10 per trip while alignments with tunnel require (and at the same time justify) toll rates of $10–$20.

Table 14: Average Toll Rate per Vehicle Required to Cover Operation and Maintenance Cost, per Trip, by Alignment and Policy Scenario, in 2030
($)

| | | West | | | | | | |
| | | New Kastek | | Masanchi | | | | |
	Scenario	Gravel	Asphalt	Gravel	Asphalt	Tunnel	Direct	East
Average Toll Rate	Scenario II	5.66	6.16	7.88	8.55	17.70	17.63	16.38
	Scenario III	2.61	2.75	4.18	4.16	7.66	7.06	6.99

Note: Toll rates are expressed in constant 2017 prices, without any profits.
Source: Consultant team analysis.

Table 15: Average Toll Rate per Vehicle Justified by Vehicle Operation Cost Savings per Trip, Corresponding to 50% of Those, in 2030
($)

| | West | | | | | | |
| | New Kastek | | Masanchi | | | | |
Item	Gravel	Asphalt	Gravel	Asphalt	Tunnel	Direct	East
Average Toll Rate	7.00	8.45	3.70	5.45	10.00	18.73	12.20

Note: Toll rates are expressed in constant 2017 prices.
Source: Consultant team analysis.

30   Two examples are (i) the Great St. Bernard tunnel between Italy and Switzerland, and (ii) the Gordie Howe International Bridge between Canada and the United States.
31   Operation and maintenance cost are expected to be 1.5% of capital investment costs (Chapter 3, section 3.5).

162.   If it seems appropriate to set the toll rates higher than that, a capital facilities charge could be added to the toll, helping to amortize parts of the capital investment. However, the toll required to do that could be seen as too expensive by travelers.

163.   If a PPP model is chosen and the alternative road becomes a toll road, some travelers may be deterred by the additional cost from traveling more frequently, especially in the cases of more costly alignments with tunnels. This would in some cases limit the economic development enabled by the alternative road.

8

Conclusions

164.   The alignments, because of their physical characteristics, fall into three groups:

(i)   The western alignments through New Kastek pass or Masanchi can be implemented for moderate capital investment costs of about $100 million as they do not include tunnels and bypass the highest mountain ridges. Travel times are 20% to 40% lower than on the existing route.

(ii)   The western and eastern alignments with tunnels require much higher capital investments of about $500 million. Travel times are approximately half as long as for the existing routes.

(iii)   The direct alignment reduces travel times by three-quarters but would require the highest capital investment costs of almost $600 million.

165.   Existing users will save from $4 million to $23 million in travel time costs and vehicle operating costs in 2030, depending on the alignment.

166.   Economic impacts as a consequence of increased economic activity enabled by the alternative road are considerable for both countries. On the average, at least $31 million to $165 million will be added each year to the national GDP of Kazakhstan depending on the alignment and policy scenario. This range will be from $53 million to $439 million per year in the Kyrgyz Republic, which corresponds to 0.8%–6.7% of the 2017 national GDP.

167.   The EIRR seems to be more independent from the policy environment for the western alignments. Alignments that include a long tunnel should presumably only be considered if scenario III is clearly preferred and other recommended policies and infrastructure improvements are implemented together with implementation of the alternative road.

168.   The financial internal rate of return, based on a single set of assumptions for the tolling scheme of a new road, suggests that some alignments require a very strong demand increase to be financially viable under these assumptions. The demand increase will depend on additional initiatives to be taken to further develop tourism in the area.

169.   An increase in travel demand is needed for any of the alignments to be economically and financially[32] viable (Table 16), which is more than likely given the considerable travel time and cost savings offered by the alternative road of any alignment. Travel forecast estimates, underpinned by qualitative information from interviews, suggest strong increases for all alignments. There appears to be latent demand for destinations within reach for short-stay trips. One of the alignments, West—New Kastek pass, manages to be economically viable with existing and induced travel demand only. All other alignments require further economic development, enabled by the road itself and supporting policies and projects, to make them viable.

---

32   Economic analysis is the clear focus of this study. Financial viability is only examined for one set of assumptions regarding costs and toll revenues.

Table 16: Economic and Financial Viability of Alignments

Demand Level	West										Direct		East	
	New Kastek				Masanchi				Tunnel					
	Gravel		Asphalt		Gravel		Asphalt		Tunnel					
	Eco	Fin	Eco	Fin	Eco	Fin	Eco	Fin	Eco	Fin	Eco	Fin	Eco	Fin
Scenario III demand														
Scenario II demand														
Existing plus induced demand														
Existing demand														

■ Viable at this demand level   ■ Not viable at this demand level   ■ Not determined

Eco = Economic, Fin = Financial
Source: Consultant team analysis.

170.   Two different basic alignment choices seem possible:

(i)   An alternative road at moderate cost (western alignments without tunnel) promises to be more independent from strong increases in travel demand, with these alignments offering limited travel time and cost savings but showing the strongest economic and financial viability.

(ii)   A more direct alternative road (especially the direct alignment) offering strong travel time and cost savings requires considerably higher capital investments and is a riskier endeavor. If supported by effective policies and projects as supposed in scenario III of the ABEC Tourism Master Plan, this choice promises stronger economic development. Firmly committed investment intentions for a ski resort like Turgen, which offers synergies, may be utilized as an additional impulse for one of the more direct alternative road alignments.

171.   A positive economic impact on the economy of the ABEC region, and both the Kazakh and the Kyrgyz national economies can be clearly stated. There are economically viable solutions that, in a supporting policy environment, enable potentially strong economic development in the region. These solutions are also financially viable if only the recovery of the cost for operation and maintenance are considered.